Saliha DJABALI

Polyphenols and resistance to fungal infestation

Saliha DJABALI

Polyphenols and resistance to fungal infestation

Effect of polyphenols on resistance to fungal infestation in dry bean grain

ScienciaScripts

Cover image: www.ingimage.com

This book is a translation from the original published under ISBN 978-620-6-70262-7.

Publisher:
Sciencia Scripts
is a trademark of
Dodo Books Indian Ocean Ltd. and OmniScriptum S.R.L publishing group

120 High Road, East Finchley, London, N2 9ED, United Kingdom
Str. Armeneasca 28/1, office 1, Chisinau MD-2012, Republic of Moldova, Europe
Printed at: see last page
ISBN: 978-620-7-79770-7

Contents

Introduction

Alongside cereals, legume grains, particularly dried beans, are a source of food and protein for humans in many regions of the world (Kellouche and Soltani, 2005). Unfortunately, in addition to yield losses in the field, beans (*Phaseolus vulgaris* L.) suffer considerable damage during storage. The main cause of this damage is mould contamination. Not only does this cause a direct reduction in dry weight, it also reduces the technological and nutritional quality of the grain. The sources of fungal contamination are numerous and grains can be contaminated by fungi during the growing season, as well as throughout the harvesting process up to storage (Bulter and Day, 1998).

In response to this contamination, chemicals have been used to inhibit fungal growth (Phattayakorn and Wanchaitanawong, 2009). However, the WHO has banned the use of certain chemical fungicides because of their undesirable long-term toxicological effects, including cancerogenicity (Chahardehi *et al.,* 2010). Similarly, increasing consumer awareness of the risks of these compounds on food products has prompted the need to look for other preservatives that are effective, natural and safe to health.

Some studies have shown that the ability of a plant species to resist attack by insects and microorganisms is often correlated with its phenolic compound content (Harborne, 1989). Substances belonging to the group of phenolic compounds, which are highly heterogeneous in terms of both composition and structure, have long been poorly understood (Lugasi *et al.*, 2003). Considered as secondary, metabolically inactive substances, they aroused little interest. Today, this view is set to change. Research in recent years has shown that phenolic compounds are by no means inert products of metabolism. In plant tissues, they are subject to major variations in quantity and quality, demonstrating an undeniable biochemical dynamic (Esekhiagbe *et al.*, 2009).

Certain hypotheses have been put forward, postulating that plants use different means to defend themselves against attacks from their environment, such as insects, bacteria, moulds, etc. Some of these means of defence involve chemical compounds which interact with the metabolism of the attacker. Some of these means of defence involve chemical compounds that interact with the metabolism of the attacker; being intoxicated and affected, the latter may be discouraged from continuing its attack. The accumulation of polyphënolic compounds could therefore be envisaged as a non-specific response to different types of aggression (Chërif *et al.*, 2007). This hvpothesis prompted us to undertake a study on the correlation between the total polyphenol content of the grains of two dry bean varieties differing in colour and resistance to fungal infestation based on the findings of Esekhiagbe *et al.* (2009).

The main objectives of this work were to isolate and identify the moulds contaminating the grains of two dry bean varieties, to assess the total polyphenol content extracted from the grains and to demonstrate their antifungal activity against the isolated strains.

To achieve these objectives, we have found it useful to structure the manuscript as follows: in addition to the introduction and the general conclusion, it is structured in three parts. The first part is a bibliographical review in which three chapters are discussed; in the first chapter, we deal mainly with the causes and consequences of fungal spoilage of dried vegetables and the means of combating it; the second chapter deals with moulds, their isolation and identification; the third chapter elucidates the main classes of phenolic compounds, their biological activities and the main methods of extracting and identifying them. This bibliographical overview has been used to support the experimental work and the interpretation of our results.

The second part presents the equipment and methods used for measuring moisture content, isolating and identifying the fungal genera contaminating the two dry bean varieties, extracting and quantifying total polyphenols from dry bean kernels and testing the sensitivity of isolated mycetes to polyphenolic extracts.

The third part presents the results obtained, followed by discussions; It summarises the evaluation of the total polyphenol content of the dry bean varieties analysed, the estimation of the percentage of contamination, the identification of the isolated mould strains, the evaluation of the antifungal power of the phenolic extracts of the varieties analysed with respect to the isolated mould strains and the demonstration of correlations between the total polyphenol content of the dry bean grains and resistance to mould infestation.

1. General

In terms of surface area and quantity produced, legumes are the second most important crop family after cereals (Gepts *et al.*, 2005). Legumes belong to the *Fabaceae* family and comprise around 20,000 species, characterised by flowers with five petals and an upper ovary that forms a pod filled with protein-rich grains (Cronk *et al.*, 2006).

Representatives of this family include: grain protein crops (peas, beans, faba beans, chickpeas and lentils), fodder protein crops (alfalfa and clover), and oleaginous protein crops (soya and groundnuts) (Cazaux, 2009). Dried legumes, which are the dried grains of protein legumes, are distinguished from oil legume grains by their low fat content (FAO and WHO, 2007).

In Algeria, the most widely grown leguminous food crops are: lentil (*Lens culinaris* L.), chickpea (*Cicer arietinum* L.), pea (*Pisum sativum* L.), faba bean (*Vicia faba* L.) and bean (*Phaseolus vulgaris* L.) (MA, 1998). These crops have received much attention from the agricultural services in order to increase the area under cultivation and improve yield levels (Table 01).

Table 01. Area and production of dried vegetables in Algeria (2006-2009) (DSASI, 2009).

	2006/2007		2007/2008		2008/2009	
	Area (ha)	Production (Qx)	Area (ha)	Production (Qx)	Area (ha)	Production (Qx)
Feve/Feverole	31284	279735	30688	235210	32278	364949
Dry peas	9184	62430	7556	36175	8487	59692
Lenses	873	5605	1309	10809	2588	26932
Chickpeas	20681	142940	20361	112110	22274	178404
Dried beans	1394	9170	1040	5441	1616	11588
Gesses	94	950	197	1980	205	1325
Total	63510	500830	61151	401725	67448	642890

Although the food legumes grown have Ьёпёйаё of some development programmes, national production of dried vegetables has not improved as expected, either in terms of area or grain production. Yields have been unstable for all species. The reasons for this situation are both technical and socio-economic (Drevon, 2009).

Worldwide, grain legumes and fodder legumes occupy almost 180 million hectares, representing 12-15% of arable land (Djebali, 2008). Grain legumes cover 33% of human dietary protein requirements. This share is provided mainly by beans, peas, chickpeas and beans (Vance *et al.*, 2000).

As well as their nutritional qualities, legumes play a vital role in preserving the environment. Nitrogen is the most limiting nutrient for plant production in most natural ecosystems. Through their symbiotic capacities, legumes can contribute to the colonisation of poorly fertile ecosystems (Sprent and Parsons, 2000).

In addition to these benefits for food and the environment, legumes can be useful in a variety of industries, including food, chemicals and pharmaceuticals (Graham and Vance, 2003).

2. Biochemical composition

From a biochemical and nutritional point of view, dried vegetables are characterised by :

> a protein content of between 20 and 25%; these proteins are mainly globulins

characterised by a lack of sulphur amino acids (methionine and cysteine);

> carbohydrates made up of perfectly digestible starch (45 to 50%) and fibre (8 to 18%); however, they also contain non-digestible a-galactosides (raffinose, stachyose, verbascose, etc.), hydrolysed and metabolised by intestinal bacteria into hydrogen, methane and other gases responsible for flatulence;

> the presence of thermolabile anti-nutritional factors (trypsin inhibitors and lectins) and phenolic compounds (Siret, 2000).

These ëlëments are distributed throughout the grain structure (Figure 01).

> **A protective envelope**: whose thickness, relative weight and rigidity vary according to the species. It is low in water and contains mainly fibres and phenolic compounds.

> **Two cotyledons**: which contain protëmes of reserves localised in protëic bodies. They also contain lipids, carbohydrates, particularly starch, vitamins and minerals.

> **The germ**: which represents 1 to 3% of the weight of the grain contains proteins (40%), lipids (10%) and carbohydrates (45%) (Cup and Legnaud-Rouand, 1992).

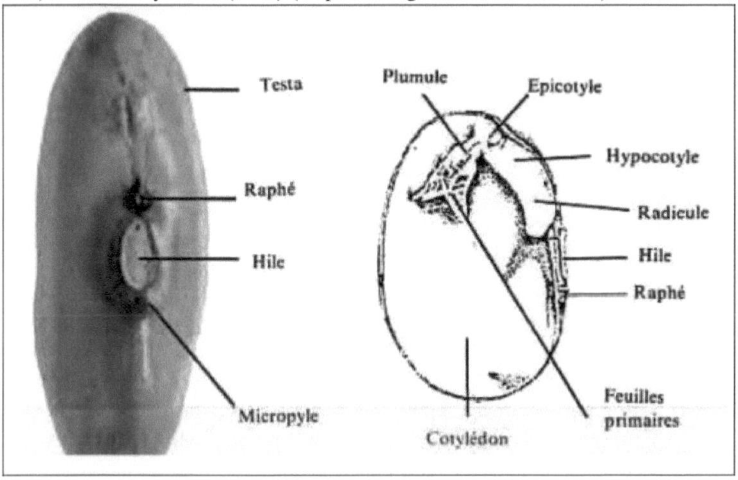

Figure 01. Structure of the dry bean (Daniel *et al.*, 1987).

Table 02 summarises the average composition of some legume grains. **Table 02:** Average composition of some legume grains (g/100g of grain) (Arkoyld and Daughty, 1982).

Grains	Water	Proteins	Lipids	Carbohydrates	Fibres	Minerals
Peas	9	23	1,5	60	8	3,5
Chickpeas	10	20	5	58	12	4
Feves	8	23,8	0,2	6	15	0,4
Beans	11	23	2	55	4	4
Lenses	11	25	1	54	17	3

3. Storage of dried vegetables

By storage we mean the phase of the post-harvest system during which products are preserved, in an appropriate manner, in order to guarantee the food security of the population outside the periods of agricultural production. In order to achieve this objective, it is obviously necessary to adopt measures aimed at preserving the quality and quantity of stored products over time (De Lucia and Assennato, 1992).

Dried vegetable grains can be stored in all kinds of containers ranging from earthen gourds, baskets and huts to large mëtal or cement silos (Hayma, 2004).

3. 1. Traditional storage

In this type of storage, the grain is kept in round or square granaries, usually made of earth with varying degrees of added vëgëtal fibre. The major drawback of this method is the very high humidity and the infiltration water, which favour the development of mould and the phenomenon of bacterial fermentation (Doumandji *et al.*, 2003).

3. 2. Storage in bags

Storage in bags in the shop is the most frequently used solution, as it requires less investment than bulk storage. Unfortunately, the bags are left on the ground and are exposed to rodents and insects that carry mould spores (Jard, 1995).

3. 3. Bulk storage

Bulk storage is still not very common in developing countries, whereas it is widespread in developed countries. Stacked grain is left in the open air under open sheds with mëtalIIque frames. Unfortunately, contamination is possible, especially as in this type of construction, there are always spices between the walls and roofs and insects are allowed to pass freely (Doumandji *et al.*, 2003).

3. 4. Storage in silos

These are cylindrical enclosures made of metal or steel, covered on the inside with a layer of aluminium to prevent condensation (Jard, 1995). These silos can be used to store several types of dried vegetable or cereal and are multi-product (Duron, 1999). This is an effective method that reduces damage and limits pest attacks (Jard, 1995).

4. Alteration factors

During storage, dried vegetables are threatened by..:

5. 1. Physico-chemical factors

6. 1. 1. humidity

Among the most important dëgradation factors, humidity plays a prëpondërant role during storage (Multon, 1982). In any mass of large grains in storage, which is not artificially ventilated, the differences in tempërature between different parts of the mass cause a displacement of the moisture and the result is often the formation of pockets of intense humidity, which constitute the ideal environment for the development of fungi (FAO, 1984). The flora of storage is composed of xerophilic species adapted to relatively dry substrates (Feillet, 2000).

7. 1. 2. Temperature

Storage temperature is the other major factor influencing grain longevity (FAO, 1994); it plays an important role in grain preservation, as it favours the phenomenon of respiration and hence the degradation of stored products. It also influences the development of microorganisms and insects (Cruz and Diop, 1989). Low temperatures slow down the activity of microorganisms on the surface of the grains, as well as that of insect larvae when they are inside the grains (Suszka *et al.,* 1994).

4.1.3 Storage duration

It is clear that the longer the storage period, the greater the loss of dry matter due simply to grain respiration. The risks of attack by various predators are also increased. For multi-annual storage, it is important to remember that the grains must be very dry and in a favourable environment to enable them to be preserved over a long period (Cruz and Diop, 1989).

4.1.4 Oxygen and carbon dioxide content

The oxygen and carbon dioxide content affects the nature of the metabolism of insects, mites, micro-organisms and grains; it also affects non-enzymatic oxidation and certain enzymatic reactions (Multon, 1982). Storing grains in a low-oxygen environment kills insects, stops the development of micro-organisms and blocks or slows down grain degradation (FAO, 1994). This principle of preservation is put into practice in techniques known as preservation under confined atmosphere or inert atmosphere (Cruz and Diop, 1989).

4.2 Biotic factors

The main causes of damage to legume storage facilities are

4. 2. 1 Insects

Insects are the main vectors of mould spores in the field and in storage (Pfohl-Leszkowicz, 2001). Most moulds grow best at temperatures of 25 to 30°C and relative humidities of 70 to 80%. At lower or higher temperatures and relative humidities, they continue to develop, albeit to a lesser degree (Hayma, 2004). There are five main species of insect that attack dry beans. These are the bean bruchid (*Acanthoscelides obtectus*), cowpea bruchid (*Callosobruchus maculatus*), Indian meal moth (*Plodia interpunctella*), tobacco moth (*Ephestia elutella*) and almond moth (*Cadra cautella*). The bean and cowpea bruchids are the most destructive (Pfohl-Leszkowicz, 2001).

4. 2. 2 Mould

Microbial biodëtërioration of grains before and after harvest is a well-known phënomëne causing significant loss of up to 30%. Mould contamination is the main damage that will lead to many problems (Satish *et al.*, 2010). Moulds or their spores are almost always present in the grain. The spores need a warm, moist environment to develop and produce hyphae. The hyphae penetrate the product and partially transform it into other substances necessary for their growth (Hayma, 2004). The genera and species making up the mycoflora of dry tegume grains are classified into three main groups.

4. 2. 2. 1 Field mycete

In the field, tegumineuses are attacked by many moulds. Field moulds include a large number of species belonging in particular to the genera *Erysiphe, Botrytis, Ascochyta, Mycosphaerella, Verticilium, Fusarium, Colletotrichum*, etc. (Djebali, 2008). They group together moulds with a phytop athogëne tendency that get involved on the grain before harvest (Godon and Loisel, 1997).

Among the mycological diseases of tegumineuses (Table 03).

Table 03. Main diseases of tegumineuses (Cazaux, 2009).

Diseases	Responsible pathogen	Culture affected
Ascochytosis	*Ascochyta spp.*	beans, lentils, chickpeas
Anthracnose	*Colletotrichum spp.* *Mycosphaerella pinodes*	beans, lentils,
Brown spots	*Botrytis spp.*	chickpea, fëve, lentil
Root rot	*Fusarium spp.* *Phytophthora spp.* *Pythium spp.* *Rhizoctonia spp.* *Sclerotinia spp.*	most grain teguminous plants

4. 2. 2. 2. storage mycete

The main moulds in storage include only a few *Aspergillus* species that are adapted to life without free water and, to a lesser extent, *Penicillium* species (Christensen *et al.*, 1982), which normally grow in environments where water activity is higher than that required for *Aspergillus* growth and at lower temperatures. These are frequent contaminants in temperate regions (Tabuc, 2007).

4. 2. 2. 3. Mycete intermediate

Is a third category with a more diversified behaviour. It includes germs capable of limited development, at the beginning of storage, in particular conditions and especially on insufficiently dry grains. This flora includes *Cladosporium, Trichoderma, Rhizopus, Absidi* and *Mucor*. Mucorales, together with the yeasts *Candida* and *Torulopsi,* are the usual representatives of this intermediate flora, the evidence of which very often reveals storage in confined and excessively humid conditions (Godon et Loisel, 1997).

4.3 Endogenous factors

4. 3. 1 Breathing

The respiratory activity of dry grain is low, due almost entirely to germ respiration. It continues after harvest and during storage, accelerating or slowing down depending on the temperature and relative humidity, as well as the moisture content of the grain. If the moisture content exceeds 14%, apparent respiration gradually increases until a critical moisture threshold is reached, beyond which respiration accelerates rapidly and the grain tends to warm up. The reducing sugars are converted into CO_2 and water, leading to a loss of sugars and dry matter (FAO, 1984).

4. 3. 2 Enzymes

the moisture content required for the activity of certain enzyme systems in the germ plays an important role in grain storage. For example, a- and P-amylases in germination-damaged grains attack the starches and convert them into maltose and dextrose. Amylase activity increases in stored grain with a high moisture content (FAO, 1984). Lipolytic enzymes, on the other hand, are the only enzymes likely to be active at low water activity levels (A_W), areas which are precisely more favourable to chemical oxidation (Thiam *et al.,* 1976).

4. 3. 3 Fermentation

When humidity exceeds 15%, carbohydrate fermentation can occur with the production of alcohol or acetic acid, causing characteristic sour odours and loss of nutritional value (FAO, 1984).

5. Mechanism of mould resistance in *Phaseolus vulgaris*

Dry beans are exposed to a wide variety of fungi (Macarde, 2004). Three types can be distinguished according to their trophic biology. Biotrophs, necrotrophs and hemibiotrophs. Biotrophic agents, such as the fungi responsible for rust or oi'dium and certain oomycetes (Albugo), need to conserve the host's tissues in order to feed and carry out their reproductive cycle, while necrotrophs, such as the fungi responsible for melting (*Pythium*) or rotting (*Botrytis),* develop by killing the tissues on which they proliferate. Finally, hemibiotrophs, such as *Magnaporthe*, *Colletotrichum* and oomycetes (*Phytophthora)* fungi, begin their cycle mainly as biotrophs, then move on to a necrotrophic phase at a given stage in their cycle (Ramiro, 2009).

Numerous studies have been carried out to understand the bean's resistance and/or defence mechanisms, in the hope of reducing the use of plant protection products. The bean has an arsenal of defences against pathogens. The first defences are physical barriers such as the protective envelope (Macarde, 2004). Induced resistance, through the accumulation of phenolic compounds and phytoalexins, as well as the activation of peroxidases, polyphenoloxidases and key enzymes in the phenylpropanoid and isoflavonoid pathways, can play a crucial role in their resistance. Cinnamic and ferulic acids are also involved. However, further work is needed to provide more evidence of the accumulation of these compounds at the right time, concentration and location, and to elucidate the regulatory genes involved in their rapid and coordinated induction in response to fungal attack (Cherif *et al.*, 2007).

Despite the existence of constitutive barriers (vëgëtal wall, antimicrobial secondary metabolites, etc.) conferring effective general resistance, the induction of defences more adapted to the infection and to each type of parasite is often necessary (Ramiro, 2009). There are, essentially, two pathways in the immune system (Jones and Dangl, 2006). Firstly, basal resistance is triggered by receptors present in the plant membrane capable of recognising molecular signals associated with microbes or pathogens (MAMPs or PAMPs for "microbial or pathogen associated molecular patterns)", such as chitin (Ausubel, 2005; Chisholm *et al.*, 2006; Numberger and Kemmerling, 2006). To bypass the basal defence, pathogens synthesise virulence effector proteins that are released in host cells (Catanzariti *et al.*, 2007). In the next stage of the evolutionary cycle, the host cells develop the means to recognise the effector proteins and respond to the attack with a robust and rapid defence mechanism. This second defence front is called effector triggered immunity (ETI) and is regulated by NB-LRR (Nucleotide Binding Leucine Rich Repeat) type proteins encoded by resistance genes (*R* genes) (Tameling and Takken, 2008). Fungal effectors are recognised by NB-LRR proteins and activate similar defence responses. ETI-type resistance, based on NB-LRR proteins, is effective against biotrophic or hemibiotrophic fungi, but not against necrotrophs, which kill plant tissues and thus prevent any local resistance response (Ramiro, 2009).

Currently, 15 specific resistance genes have been genetically characterised in common bean. Eight resistance genes are specific to anthracnose (Ramiro, 2009).

In vitro and *in vivo* trials using different *Phoma* strains suggest the involvement of two enzymes, chitinase and glucanase, in the ascochytosis resistance mechanism (Baudoin *et al.*, 2011).

6. Varietal improvement of *Phaseolus vulgaris* dry beans

Varietal improvement has long received particular attention. The improvement objective is a combination of production potential, adaptation to different agro-ecological zones, tolerance to biotic and abiotic stresses and technological quality (Feliachi, 2006).

Among the species used to improve the common bean, *Phaseolus polyanthus* takes first place through interspecific hybridisation, as it has a high level of tolerance to biotic and abiotic constraints (Baudoin *et al.*, 2011).

A limited number of amëtëliorëes were introduced, including *Mexican 142, Red Wolaita, Awash, Melke, Beshbesh* and *ROBA-1*, which are resistant to angular bean spot, common bacterial blight, rust, anthracnose and bean fly (CIAT, 2003).

In Algeria, the obstacles limiting the availability of new varieties can be summarised as follows:

> there are no establishments specialising in plant breeding;

> slowness in approving and marketing new varieties;

> reluctance on the part of seed-producing establishments to take on new varieties, for fear of not being able to sell them on the market, given that Algerian farmers have difficulty adopting new varieties due to inadequate extension systems;

> low contribution of research in the fields of breeding and seed production (Feliachi, 2006).

7. Consequences of alterations on the biochemical composition and market value of dried vegetables

Moulds are aggressive and degrading only in their mycelial form. In their spore form, they can disperse very widely and contaminate, but are inert as long as the environment does not allow them to develop. It is therefore important to distinguish between contamination by spores and degradation, which is the active phase caused by expanding mycelium (Roquebert, 1997).

Moulds significantly modify the substrate on which they live, so mould growth on grains is accompanied by a change in their appearance, odour and taste, a loss of dry matter, a reduction in nutritional value and a change in their composition (Cahagnier, 1996).

7. 1. Degradation of carbohydrates

Dried vegetables contain around 60% carbohydrates, most of which are represented by starch. The degradation of starch is one of the most common processes and involves several types of enzyme: -amylase, which has an endomoietic action leading to the formation of *D-glucose*, maltose and a small amount of maltodextrin, gluco-amylase, which releases glucose units from the non-reducing ends of polymers, and f3-amylase, which has an exomolecular action, producing maltose (Guiraud, 2003).

7. 2. Lipid degradation

Grain lipids, particularly triglycerides, are particularly sensitive to degradation by microorganisms. The first biochemical changes that grains undergo during poorly-conducted storage are most often observed in the lipids (Feillet, 2000). Triglycerides are hydrolysed into glycerol and fatty acids using lipases, which are found in moulds of the genera *Rhizopus, Aspergillus* and *Geotrichum* (Guiraud, 2003).

7. 3. Degradation of proteins

Some dried vegetables, such as beans, are rich in protein. Thanks to their proteolytic enzymes (protease), moulds hydrolyse proteins, which leads to the formation of amino acids that are easily used by moulds. There are also hydrolysis substances such as hydrogen sulphide ($H_2 S$) and ammonia. This degradation essentially causes the emission of foul smells and a profound change in consistency and taste, which becomes bitter. The main species that cause this hydrolysis are: *Aspergillus flavus, Aspergillus parasiticus* and *Aspergillus niger* (Multon, 1982).

7. 4. Biosynthesis of mycotoxins

Mycotoxins are low molecular weight secondary metabolites produced by various moulds under certain environmental conditions. They are found in the mycelium and spores and can diffuse into the substrate (Brochard and Le Bacle, 2009). A given mycotoxin is not necessarily specific to a single mould; similarly, a given mould may produce several different toxins (D'halewyn *et al.*, 2002). The chemical origin of mycotoxins is highly diverse, some deriving from amino acids (ergot alkaloids, aspergillic acid, etc.), polycetoacids (Aflatoxins, Ochratoxin, Patulin, etc.), terpenic dërivës (Diacëtoxyscirpënol, Fusarenone, etc.) or fatty acid dërivës (Fumonisins, Alternariol) (Pfohl-Leszkowicz, 1999).

According to FAO (1984), around 25% of foodstuffs are contaminated with mycotoxins, which pose several types of risk:

> **Toxicological**: Toxicity; increased prevalence of certain cancers and allergies, as well as very serious damage to the kidneys, liver, genes, immune system, genital tract and nervous system;

> **Technological**: enzyme inhibition, resistance to treatment;

> **Economic**: The poorer performance of animals fed contaminated feed leads to rejection by consumers.

According to Pfohl-Leszkowicz (1999), the main types of mycotoxin found in dry beans are: aflatoxins, ochratoxin, penicillic acid and sterigmatocystin.

8. Prevention and control methods

Mould can be controlled before it attacks the grain. This is known as preventive control. Sometimes it is necessary to combat mould directly, and this is called curative control.

8. 1. Preventive action

This control involves rigorous hygiene of harvesting machines, handling facilities, means of transport and storage premises. It is important to isolate new harvests from old harvests in the warehouse (Kellouche and Soltani, 2005) and, in order to prevent or at least limit the emigration of spore-carrying insects to crops, the premises in which the dried vegetables are stored should be cleaned and, if necessary, disinfected. In cultivation, the use of resistant varieties can be an effective method of control (Goix, 1986).

8. 2. Curative control

8. 2.1 Chemical control

The most significant damage occurred during the storage period. Several types of fungicide were used, including low-weight organic acids.

(propionic acid, acëtic acid and formic acid) and their salts are the most widely used (Magan and Olsen, 2004). However, some mycëtes can use these acids as a carbon source, hence the need to look for alternative grain preservation methods (Tatsadjieu *et al.*, 2009). In the case of large storage lots, fumigation is an effective technique. Unfortunately, however, among the chemical substances used, methyl bromide and ethylene oxide cause the induction of phënomënes of resistance in toxinogenic genera (Flamini and Luigi Cioni, 2003).

8. 2. 2. Physical combat

Physical agents such as temperature or radiation have proved to be very effective (Botton *et al.*, 1990). The lethal action of gamma irradiation on living organisms results from chemical modifications, even quantitatively minute, induced in their vital molecules. Toxin-producing fungi are more sensitive to irradiation than others. However, many practical factors, both technical and biological, continue to limit the use of gamma irradiation (Magan and Olsen, 2004).

8. 2. 3 Biological control

The intensive and indiscriminate use of chemical fungicides in storage facilities has led to contamination of the biosphere and the food chain, eradication of non-target species such as auxiliary fauna and the appearance of resistant micro-organisms (Magan and Olsen, 2004). Biological control has therefore become essential. The mechanisms by which fungal growth and mycotoxin production can be prevented in the competitive environment include competition for nutrients and space and the induction of defence mechanisms. Although biological control does not yet appear to be feasible on grains, an important benefit that could emerge from studies on antagonistic mycetes would be the discovery of compounds that inhibit mycete and/or storage mycotoxin production (Magan and Olsen, 2004).

8. 2.4 Control using natural products

Recently, pesticide residues on horticultural produce, particularly fungicides used at the post-harvest stage, increased consumer awareness of the risks of these residues on food products and their effect on human health, and the emergence of mycete resistance to certain chemical fungicides, are forcing farmers to look for new substances and preservation methods, including the use of natural products such as essential oils and polyphenolic plant extracts (Shirzadl *et al,* 2011). However, contradictory results have ëlë reported in many cases, since several parameters must be considered when these results are compared: the growing medium, the botanical source of the plant material and the extraction techniques (Lis-Balchin, 2002).

Chapter 02

Mould

1. General characteristics

Moulds are filamentous eukaryotic organisms (Reboux *et al.*, 2010). Some live in symbiosis with vëgëtals, others are parasites of vëgëtals and/or animals, others are saprophytes, growing on organic dëchets (Bouchet, 2005).

This is a hëtërogëne group of microscopic, ubiquitous fungi, caractërisës by: ❖ the chemical nature of the cell wall, consisting essentially of stërol (ergostërol) and polysaccharides (P-glucans and chitin) (Chabasse *et al.*, 2002);

❖ reproduction by sexual or asexual spores (Tabuc, 2007);

❖ the presence of ^увсдёно, as a reserve substance (Tabuc, 2007);

❖ the absence of chlorophyll, i.e. heterotrophs (Tabuc, 2007).

They are found everywhere in nature and have two facets, l'a tenific: *Penicillium camembertii* and *Penicillium roquefortii* in cheese-making; *Penicillium jensenii* or *nalgiovense* in cured meats (Boudra, 2002), use in medicine for the synthesis of antibiotics and use in agriculture as a means of combating irematodes (Galtier *et al.*, 2005). It should also be noted that the biomass produced by *Penicillium chrysogenum* is used as fertiliser or feed for ЬёlяП (Nguyen, 2007). The other harmful facet is the quantitative and qualitative reduction in the food value of the commodity, a drop in crop yield and serious health probes (Krogh, 1987).

Two groups of toxinogenic fungi can be distinguished: the first type consists of fungi that invade their substrate and produce the mycotoxin on senescent or stressed plants: these are known as field toxins (Brochard and Le Bacle, 2009). The other group is made up of fungi that produce toxins after the harvest, and are referred to as storage toxins (Galtier *et al.*, 2005).

2. Classification

The classification of moulds is essentially based on purely morphological criteria (Meyer *et al.*, 2004). They are a veritable kingdom ëequalling in rank that of plants and animals. They share with animals the ability to export the hydrolytic enzymes that break down the polymëres used in nutrition (Hawksworth *et al.*, 1995) and with vëgëtals the vëtative apparatus 'thallus' dëpourvu of stems, roots and leaves (Taylor, 1993). The major subdivisions characterising moulds are summarised in Table 04.

Table 04. Classification of moulds (Meyer *et al.*, 2004).

Subdivisions	Classes	Orders	Genres
Zygomycotina	*Zygomycetes*	*Mucorales*	*Rhizopus*
Ascomycotina	*Plectomycetes* *Pyenomycetes* *Hemiascomycetes*	*Eurotiales* *Spahaeriales* *Endomycetales (yeasts)*	*Emericella* *Neurospora* *Alternaria* *Fusarium* *Eremothecium* *Puccinia*
Basidiomycotina	*Hemibasidiomycetes*	*Urenidales* *Ustilaginales*	*Aspergillus* *Moniliella*
Deuteromycotina	*Hyphomycetes*	*Moniliales*	*Penicillium*

3. Development mode and conditions

4. 1. Mode of development

Colonisation of the substrate is achieved by extension and branching of the hyphae. The apical regions of the hyphae are the active parts where most of the synthesis and degradation reactions of primary metabolism take place. They are characterised by the presence of numerous cytoplasmic vësicles containing enzymes and precursors for the synthesis of new polymers. The products of secondary metabolism tend to be stored in the subapical region, the best known of which are pigments, antibiotics and mycotoxins. The hyphae are applied to the substrate or sometimes immersed in it. They absorb water, nutrients and ions through their walls (Roqubert, 1997).

5. 2. Development conditions

Like all microorganisms, fungal growth is also dependent on a number of factors:

3.2.1 Nutrient factors

The most important nutrients are carbon and nitrogen (Abdel Massih, 2007).

Glucose, fructose, mannose, galactose, maltose, sucrose, starch and cellulose represent the sugars most ий^ёз by moulds as a source of carbon and energy. These carbohydrates are dëgradës thanks to glycolysis and aëroЬIe mëtabolism (Boiron, 1996; Nicklin *et al.*, 2000).

Nitrogen is the second most important chemical element in the cell material, consisting mainly of protëines and nucleic acids. It accounts for an average of 12% of cell dry weight (Scriban, 1993); it cannot be assimilated in the form of atmosphëric nitrogen by moulds. The latter metabolise nitrate, ammonium, urea and amiic acids, which are more easily assimilated (Botton *et al.,* 1990). Peptides and protëines are only used by moulds after their degradation (Botton *et al.*, 1990; Nicklin *et al.*, 2000).

Copper, mag^sium, sodium, zinc and molybdëne are micronutrients (Nicklin *et al.*, 2000) that are required by most fungal species for the production of cytochromes, pigments, organic acids, etc. In addition, phosphorus, potassium, calcium and sulphur represent macronutrients required by moulds and are converted into various compounds after reduction (Boiron, 2000). In addition, phosphorus, potassium, calcium and sulphur represent the macronutrients required by moulds and are converted into various composës after reduction (Boiron, 1996).

Various vitamins are also important for their growth, in particular thiamine and biotin, which act as coenzymes during carboxylation (Riviere, 1975; Botton *et al.*, 1990).

3. 2. 2 Physico-chemical factors

3. 2. 2. 1 Water activity (Aw)

Moulds are more xërotolërantes than other microorganisms (Ibicreries, yeasts). Most of them grow well at Aw values of around 0.85, whereas moulds belonging to the genera *Aspergillus* and *Penicillium* are generally capable of growing at Aw values of around 0.7 at 25°C; they can therefore develop on water-poor foods such as cereals during storage, dried vegetables and products with reduced water activity (dried cured meats, jams, etc.).). By comparison, *Fusarium* can only develop at Aw levels above 0.9. These are therefore species that develop in the field, on living plants (Castegnaro and Pfohl-Leszkowicz, 2002).

3. 2. 2. 2. pH

Micromycëtes can grow in a wide pH range; they grow for pH between 3 and 8, with optimal growth between 5 and 6 (Keller *et al.*, 1997). Some species such as *Aspergillus niger* can grow up to 1.7. This ability to grow in an acidic environment means that they can be separated from bacteria for isolation (Larpent et Larpent, 1990).

3. 2. 2. 3. Temperature

Moulds are generally mesophilic, and hyphal growth is optimal between 20 and 25°C. Outside this temperature range, hyphae grow more slowly. There are also psychrophilic species, such as *Penicillium expansum, P. verrucosum and P. viridicatum*. They can develop at temperatures below 4°C. These species are responsible for spoilage of cold-preserved foods (Pfohl-Leszcowicz, 2001). Heat-tolerant species are rarer. This is the case for *Aspergillus flavus*. The optimum temperature for its growth is between 25 and 35°C, but this mould can grow well in a wider range (15-45°C) and sometimes up to 50°C (Castegnaro and Pfohl-Leszkowicz, 2002). It should be emphasised that the vast majority of moulds are sensitive to a rise in temperature, and a cooking or pasteurisation treatment usually destroys the mycelia as well as the fungal spores (Pfohl-Leszkowicz, 2001).

3. 2. 2. 4 Presence of oxygen

Moulds are aerobic micro-organisms. However, their development is little affected by levels 10 times lower (2.1%) than those in the atmosphere. As a result, certain species of mould will be able to grow on foodstuffs stored in an oxygen-poor atmosphere. For example, *Aspergillus flavus* can grow in the presence of high concentrations of nitrogen 'N_2' (99%) or low concentrations of oxygen 'O_2' (0.5%), but a concentration of 80% carbon dioxide 'CO_2 ' inhibits its growth. Similarly, the development of *Aperg^ilUis ochraceus* is completely inhibited by 80% CO_2 (Pfohl-Leszcowicz, 2001). *Penicillium*
verrucosum can tolerate CO_2 concentrations of up to 25% without any significant change in its development; above 25% CO_2, the growth of the fungus is reduced by 40% and by almost 75% in an atmosphere containing 50% CO_2 (Cairns-Fuller *et al.*, 2005). On the other hand, *Fusarium proliferatum* is dependent on the presence of oxygen (Keller *et al.*, 1997) and the absence of O_2 considerably reduces fungal biomass. Some species can develop anaerobically, as in the case of *Byssochlamys*, which contaminates fruit juices preserved by pasteurisation (Pfohl-Leszkowicz, 2001).

3. 2. 2. 5. light

Light promotes the maturation and germination of spores. Moulds are generally indifferent to the action of light. However, some species (*Tuberales*) cannot tolerate light and grow in dark places (caves). Conversely, others thrive on permanently sunny mountain slopes or in desert regions (*Discomycetes*) (Pfohl-Leszkowicz, 2001).

3. 2. 3 Biological factors

3. 2. 3. 1 Presence of insects

Insects, mites and weevils are the main vectors of mould spores, which they introduce into the grain through the lesions they create. Insect infestation predisposes the grain to mould contamination and mycotoxin production (Brochard and Le Bacle, 2009).

Analysis of the fungal spores present on the surface of the insects' bodies and in their intestines revealed the presence of spores of *Aspergillus flavus, Penicillium brevicompactum, P. chrysogenum, P. cyclopium, P. griseofulvum*, *Botrytis cinerea, Cladosporium herbarum* and *Alternaria alternata*. These different fungal species were then found in the samples analysed (Hubert *et al.*, 2004). This shows that insects are far from playing a minor role in contamination in the field or in storage areas.

3. 2. 3. 2 Microbial interactions

Competition for nutrients and space is a frequently encountered phenomenon in the living world. The simultaneous presence of several species of microorganisms in the same environment leads to interactions between the different species. Environmental conditions can favour certain species and disadvantage others. For example, *Mucoraceae, with their* high apical growth rate, rapidly invade the environment, inhibiting the development of slower-growing species. The synthesis of toxic substances (mycotoxins) and their accumulation in the environment can also have an inhibiting effect on the development of other fungal species (Pfohl-Leszkowicz, 2001).

4. Reproduction

Fungi reproduce in two ways: sexually by the fusion of two gametic cells and asexually or vegetatively (Champion, 1997).

4. 1. Sexual reproduction

Moulds reproduce according to a sexual reproductive cycle (Bousseboua, 2005) which involves the production of sexual organs and gametes, the fusion of gametes or organs followed by karyogamy and meiosis and the development of fruiting bodies (Larpent and Larpent, 1990). This type of reproduction allows hereditary characteristics to be recombined. Sexuality exists only in certain species, in others it has never been demonstrated, and in others it is inconstant or rare (Guiraud, 2003).

4. 2. Vegetative reproduction

It is much more widespread than the former, and although it does not involve genetic transformation, it plays a major role in the dissemination of species. It occurs either by fragmentation of the thallus or by the production of asexual spores (Branger *et al.*, 2007). In moulds, several types of vegetative reproduction spores can be produced (Branger *et al.*, 2007): arthospores (from cell wall septation), sporangiospores (formed in a sporangium at the tip of a hypha), conidiospores (produced at the periphery of hyphae) and blastospores (formed by surface budding of a vegetative cell) (Bousseboua, 2005).

5. Isolation

Most natural environments (air, soil, water, food raw materials and manufactured products) contain microorganisms. Virtually no environment can resist their colonisation (Moreau, 1996). The identification and isolation of moulds on a contaminated substrate are carried out on culture media designed to eliminate bacteria and, if necessary, to specifically promote certain contaminants (Cahagnier *et al.*, 1998). Isolation media can be classified into three categories:

5. 1. Culture media

The culture media are not very selective, allowing the isolation of a large number of moulds, Czapek medium allows the elimination of bacteria due to its acidity, oxytetracycline glucose agar (OGA) medium and Sabouraud medium are used for the isolation of yeasts and moulds (Botton *et al.*, 1990). In the case of phytopathogenic species, it is preferable to use media based on plant decoctions such as potato gelose medium and bean gelose broth (Guiraud, 2003).

5. 2. Selective media

These media are adapted to the search for a species or group of species with a particular ecology that is difficult to detect using an ordinary medium. Antibiotics are added to these media to inhibit the growth of bacteria undesirable in mould isolation; the antibiotics used include penicillin (5-20ppm), gentamicin (50-60ppm), chloramphenicol (50-100ppm), etc. The use of media to inhibit invasive moulds is necessary in the case of certain fungi such as *Mucor* and *Rhizopus*, which rapidly invade isolation media and thus prevent the expression of other fungal species. Jarvis medium and dichloran rose bengal chlortetracycline (DRBC) medium can be used for the selective isolation and enumeration of food moulds (Botton *et al.*, 1990).

5. 3. Differential media

These media are used to determine fungi that most often belong to genera that are difficult to

identify, by encouraging the growth of certain genera or species, thus facilitating their characterisation; for example, *Aspergillus flavus* and *Aspergillus parasitacus* isolated on *Aspergillus flavus* et *parasiticus* agar medium (AFPA), which allows simple characterisation by the yellow-orange colour of the back of the colonies, creatine gelose medium (CREA), which allows various species to be identified by their appearance.
of *Aspergillus, Penicillium,* and *Fusarium* on gelose with Dichloran-chloranphënicol (DCPA) or on gelose with iprodione and dichloran (CZID) (Guiraud, 2003).

6. Mould identification

Identifying the large number of fungal species that can colonise foodstuffs and alter their quality, or even produce mycotoxins, is an essential step in reassessing the mycological risk. For a long time, this identification *was* based exclusively on observing the cultural and morphological characteristics of the species. Recent advances in molecular biology have made it possible to propose tools to aid identification (Tabuc, 2007).

7. 1. Morphological identification

The identification of a fungal species is based on the analysis of cultural criteria (temperature and growth rate, favourable environments) and morphological criteria. The latter are macroscopic (appearance of colonies, their underside, etc.) and microscopic (appearance of mycelium, spores, phialides, etc.) (Cahagnier et Molard, 1998).

7.1. 1. Macroscopic identification criteria

Filamentous fungi form fluffy, woolly, cottony, velvety, powdery or granular colonies; sometimes they can have a hairless appearance, flat or wrinkled relief and vary in consistency and size. The most common colony colours are white, cream, yellow, orange, red to violet or blue, green, brown to black. Pigments can be localised in the mycelium (*Aspergillus, Penicillium*) or diffuse in the culture medium (*Fusarium*) (Botton *et al.*, 1990).

6. 1. 2. Microscopic identification criteria

Microscopic examination of a fungal colony is carried out after spreading it between a slide and a coverslip. Generally, examination with an objective of 40 is sufficient to reveal most of the important elements of the diagnosis (Cahagnier et Mollard, 1998).

All fungi have a vegetative apparatus made up of filaments (hyphae) which together form the filamentous thallus or mycelium; the thallus may be siphonous or septate (Badillet *et al.,* 1987), and the spores may be endogenous or exogenous, formed thallic or blastically (Figure 02) (Botton *et al.*, 1990).

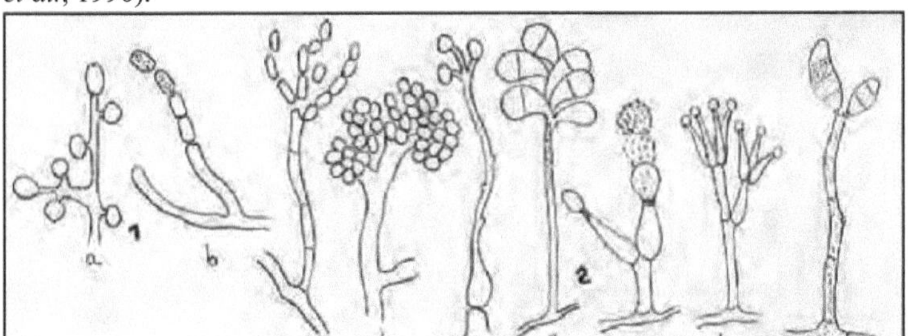

Figure 02. Conidia formation (Botton *et al.*, 1990).

1. Thallic formation: a: solitary (*Chrysogenum*), b: arthritic (*Geotrichum*)
2. Blastic formation: c: acropete (*Cladosporium*), d: sinchrone (*Botrytis*), e: sympodial (*Beauveria*), f: regressive

(*Trichothecium*), g: annelidic (*Scopulariopsis*), h: phialidic (*Penicillium*), i: poric (*Curvularia*).

Five groups of spores can be distinguished on the basis of shape and septation patterns: Amërospores (*Penicillium* and *Aspergillus*), Didymospores (*Trichothecium*), Phragmospores (*Curvularia*), Dictyospores (*Alternaria*), Scolecospores (*Fusarium*), regroupëes at the extremity of the conidiogenic cell in clusters (*Beauveria, Trichothecium*), in masses (*Botrytis*), in heads (*Acremonium, Trichoderma*), in basipetal chains (*Scopulariopsis, Aspergillus, Penicillium*) or in acropetal chains (*Cladosporium, Alternaria*) (Botton *et al.*, 1990).
Conidiogenous cells can arise from more or less elaborate structures derived from the тусёНит vëgëtative. This is ий^ё for the identification of genera and espëces. They can be directly inserted on vegetative filaments (*Acremonium, Fusarium*); well distinct from vegetative filaments borne by conidiophores dispersed on the vegetative thallus; well distinct from vegetative filaments borne by group conidiophores (Tabuc, 2007).
The presence of protective structures resulting from asexual or sexual reproduction and the presence of chlamydospores constitute another microscopic identification element (Tabuc, 2007).

***6. 2.* genetic identification**

The identification of fungal species, traditionally based on macroscopic and microscopic cultural and morphological characteristics, generally requires several days of culture (usually 7 to 10 days). Culture on specific media may be necessary to obtain conidial formation and, in some cases, the absence of conidial appearance will make it impossible to identify the mycelium. Consequently, many studies have aimed to develop identification tool methods based on the study of nucleic acids (DNA and RNA) and no longer necessarily requiring morphological examination (Peterson, 2006).
The most interesting methods are based on PCR (polymerase chain reaction) amplification of specific regions such as the gene encoding the 18S ribosomal subunit (D1-D2 region) and the internal transcribed spacers 1 and 2 (ITS1 and ITS2) regions (Hinrikson *et al.*, 2005).
Oligonucleotide probes directed to the ITS2 region of the ribosomal DNA of *Aspergillus flavus, A. fumigatus, A. nidulans, A. niger, A. terreus, A. ustus,* and *A. versicolor* allowed the differentiation of 41 isolates and gave no false positive reaction with 33 species of *Acremonium, Exophiala, Candida, Fusarium, Mucor, Paecilomyces, Penicillium, Rhizopus*, or other *Aspergillus* species (De Aguire *et al.*, 2004). This method is also used to differentiate and identify moulds responsible for food spoilage, mainly *Penicillium* species (Hageskal *et al.*, 2006).

Chapter 03

Polyphenolic compounds

1. General information on polyphenols

The term polyphënol was ële introduced in 1980, replacing the old term tannin vëgëtal and has ëlë dë defined as follows: Water-soluble phënolic compounds with a molecular weight of between 500 and 3000 Dalton and having, in addition to the usual propertiesëtës of phënols, the ability^ to precipitate alkaloi'des and дё^-йис (Cowan, 1999).

They are secondary plant metabolites (Garcia-Salas *et al.*, 2010) ëlaborës through the shikimate pathway and characterisedërisës by the presence of an aromatic ring bearing free hydroxyl groups or engagedës with a carbohydrate (Charpentier and Boizot, 2006).

These phytonutrients are responsible for pigmentation (leaf colour, fruit and flower colour) (Serrano *et al.*, 2010) and also play a role in plant growth, reproduction and protection against pathogenic aggression (Drewnoski *et al.*, 2000; Zem and Fernandez, 2005).

They are further subdivided into simple phënols (C_6 : hydroquinone), phënolic acids (C6-C4: *p-hydroxybenzoic* acid), coumarins (C3-C6: *p-coumaric* acid), naphthoquinones (C6-C4 : juglone), stillrenoids (C6-C2-C6: trans- resvëratrol), flavonoids (C_6 - C_3 -C_6 : kaempfërol), isoflavonoids (daidzëine), and in anthocyanins (delphinidol). The polymërisëes are lignans and condensed tannins (Garcia-Salas *et al.*, 2010).

2. Main classes of phenolic compounds

The authors maintain that several thousand polyphënols have ëlë already been identified and that several hundred are present in the edible parts of foods (Table 05). The most common catëgories of polyphënols are phënolic acids, stilbenes, lignans and flavonoids (Charles and Benbrook, 2005). They form the largest group of phytochemical composës in plants (Beta *et al.*, 2005).

2.1 Phenolic acids (C6-C1 or C6-C3)

Are contained in a number of agricultural and medicinal plants (Psotova *et al.*, 2003). Phënolic acids include cafëic acid, vanillic acid, fërulic acid and gallic acid (Hale, 2003). They are considered to be phytochemicals with antioxidant, chelating and anti-inflammatory effects. Their toxicity is low and they are generally considered non-toxic (Psotova *et al.*, 2003).

2. 2. Lignans (C6-C3)2

Flaxseed is an important dietary source of lignans, although these compounds are also found in smaller quantities in several other cereals, legumes and vegetables (Charles and Benbrook, 2005).

2. 3 Stilbenes (C6-C2-C6)

They are only found in small quantities in the human diet. Resveratrol is the most commonly studied polyphenol in this category (Manach *et al.*, 2004).

2. 4 Flavonoides (C6-C3-C6)

Most flavonoids have a similar chemical structure: two aromatic rings linked by three carbon atoms to form a heterocyclic oxygen compound. Flavonoids are divided into six subcategories: flavonols, flavones, isoflavones, flavanols (catechins and proanthocyanidins), flavanones and anthocyanidins (Charles and Benbrook, 2005).

Table 05. Classification of families of phënolic composës (Garcia-Salas *et al.,* 2010).

Carbon number	Class	Chemical structure	Sources

C6	Simple Phënols	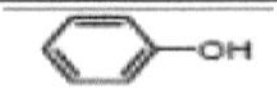	Cereals, apricots, bananas, cauliflower
	Benzoquinones	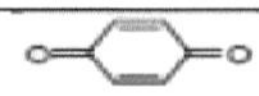	
C6-C1	Benzoic acid	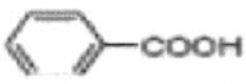	
C6-C2	Acëtophënones	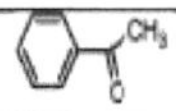	
	Phënylacëtic acid	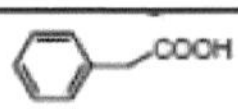	
C6-C3	Cinnamic acid	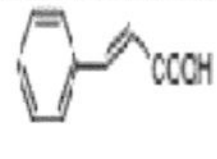	Carrot, tomato, cereal, aubergine
	Coumarins	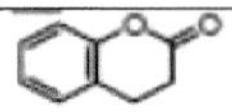	Carrot, celery, lemon, parsley
C6-C4	Naphthoquinones	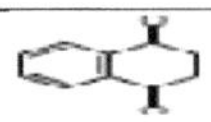	Apricot
C6-C1-C6	Xanthones	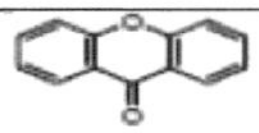	Mango
C6-C2-C6	Stilbenes		Grape
	Anthraquinones		
C6-C3-C6	Flavonoides	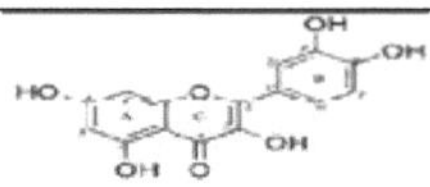	Widely distributed
$(C6\text{-}C3)_2$	Lignanes	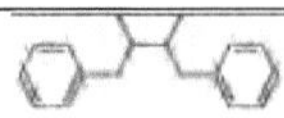	Rye, wheat
$(C6\text{-}C1)_n$	Hydrolysable tannins	Polymereheterogene composed of phenolic acids and simple sugars	Pomegranate, raspberry
$(C6\text{-}C3)_n$	Lignins	Polymerearomatic strongly reticulated	

***3.* Biological activities of polyphenols**

Polyphënols are a large chemical class. They have an extreme variëtë of structures and biological activities (Queiroz-Monici *et al.*, 2005).

3.1 Antioxidant activity

Studies on polyphënols are currently booming. Many of these studies have been carried out to inform consumers and public authorities about the benefits of fruit and vegetables rich in polyphënols, natural antioxidants with high antioxidant potential (Bouayed *et al.*, 2007).

Antioxidant capacity is the main physiological role attributed to polyphënols (Navarro *et al.*, 2008). Halliwell and Gutteridge (1995) define an antioxidant as 'a substance which, at low concentration relative to that of an oxidabic substrate, significantly retards or inhibits its oxidation'. The antioxidant action of a phenolic compound may result from a combination of chemical events, including enzymatic inhibition, metal chelation or hydrogen donation and oxidation to a stable radical (Parr and Bolwell, 2000).

Antioxidant capacity can be assessed as the potential to trap free radicals, by directly measuring radical inhibition when the antioxidant compound is added. Evaluation tests use 2,2-diphenyl-1-picrylhydrazyl (DPPH), a stable coloured radical, or 2,2'-azino-bis-(3-ethylbenzthialozine-6-sulphonic acid) (ABTS), whose coloured radical can be generated by enzymatic reaction (Rice-Evans *et al.*, 1995). The decolourisation of the radical after addition of the antioxidant is then measured (Parr and Bolwell, 2000).

3.2 Antimicrobial activity

Polyphenols are the main antimicrobial compounds in plants, with diverse modes of action and inhibitory and inhibitory activities against a wide range of prokaryotic and eukaryotic microorganisms (bacteria and fungi) (Cowan, 1999).

The mechanism behind the antimicrobial effects of polyphenols is undoubtedly highly complex (Hadi, 2004). Table 06 sets out some of the hypotheses put forward by Cowan (1999).

Table 06. Antimicrobial modes of action of some polyphënols (Cowan, 1999).

Different phenolic compounds	Examples	Mechanism
Simple phenols	Catechol Epicatechine	Substrate deprivation. Interruption of membrane function.
Phenolic acids	Cinnamic acid	Destruction of the cell wall and
Quinones	Hypericine	deactivation of enzymes.
Tannins	Ellagitannin	Binding to proteins. Enzyme inhibition. Substrate deprivation. Complex with the cell wall. Interruption of membrane function. Complexes with metal ions.
Coumarins	Warfarin	Interaction with eukaryotic DNA (antiviral activity).

3. 2. 1 Antibacterial activity

Among the antibacterial polyphënols, flavonoids block the synthesis of *Escherichia coli* nucleic acids. They inhibit the various functions of the cytoplasmic membrane by reducing the fluidity of the inner and outer layers. These compounds have bactericidal and bacteriostatic activity, disrupting energy metabolism (Jones *et al.*, 1994).

3. 2. 2 Antifungal activity

Most polyphenols have very powerful antifungal activity. Orturno (2005) demonstrated the activity of flavanone glycosides and polymethoxyflavones from *Citrusparasidi* and *Citrus sinensis* on *Penicillium digitatum*. Also, the flavonoids in *Conyza aegyptica L.* have fungicidal and fungistatic action on various agents of mycosis: *Microsporum canis, M. gypseum, Trichophyton mentagrophytes, Candida zeylanoides* (Batawita, 2002).

Table 07 lists some of the properties attributed to polyphenols.

Table 07. Biological activities of some polyphënolic composës.

Polyphenols	Activities	Authors
Phenolic acids (cinnamic and benzoic)	Antibacterial Antifungal Antioxidant	(Cowan, 1999). (Lattanzio *et al.*, 2001) (Bouayed *et al.*, 2007).

Proanthocyanidins	Antifungal Antioxidant	(Bahorun *et al.*, 1996). (Brownlee *et al.*, 1992).
Gallic and catechic tannins	Antioxidant	(Macheix *et al.*, 2005). (Macheix *et al.*, 2005).

4. Variability factors in polyphenol content

4.1 Effect of external factors

Phenolic metabolism is particularly sensitive to the action of external factors such as light, temperature, pathogenic microorganisms and human treatments (Dinelli *et al.*, 2006).

4. 1. 1. light

The importance of light (visible spectrum but also UV contained in solar radiation) on the accumulation of anthocyanins is summarised by the intervention of two paraтёйсз: on the one hand the intensity of the luminous flux and on the other the nature of the constituent radiations. It acts directly, through the intermēdiary of blue and red radiation and the plant pigment phytochrome on the activation of the promoter of the phenylalanine ammonia lyase (*PAL*) genes, which then results in the transcription of mRNAs and then the formation of the enzymatic protein (metabolism enzyme) (Hahlbrock *et al.*, 1995).

4. 1. 2. Temperature

Temperature is also a factor in regulating the expression of phenolic metabolism, often in interaction with light. Thus, a reduction in temperature associated with adequate light treatment frequently induces an accumulation of anthocyanins. Here again, regulation could be at the level of *PAL* itself, with inhibitors of the enzyme being set up under the effect of high temperatures. Disturbances in phenolic metabolism can sometimes occur as a result of cold treatment of vëgëtal organs, leading to browning (Rhodes *et al.,* 1981).

4. 1. 3 Pathogenic microorganisms

The contamination of vëgëtal by pathogenic micro-organisms also leads to a sharp increase in the levels of phënolic compounds, corresponding to the development of plant defence mechanisms (Dixon and Paiva, 1995).

4. 1. 4 Treatments applied by man

Certain treatments (application of fertilisers, irradiation, etc.) can modulate the plant's content of phënolic compounds, either during growth or during conservation of the plant organs. The consëquences are often predictable because the response can vary greatly from one species to another and according to the doses applied and the duration of treatments. Profound changes in phenolic equipment also occur when plant organs are subjected to technological processes designed to transform them (bleaching, cooking, etc.) (Macheix *et al.*, 2005).

4.2 Physiological stage

The phenolic compound content of plant organs also varies according to physiological stage. With the exception of anthocyanins, the concentration of phenolic compounds decreases during growth and ripening. Each group of phenolic compounds can evolve during growth according to its own kinetics, leading to varying proportions of the different compounds depending on the physiological stage reached (Macheix *et al.*, 2005).

4.3 Effect of species and variety

The expression of phenolic metabolism in the plant is the translation of the genetic heritage specific to each species. The nature and content of accumulated phenolic compounds are therefore primarily a characteristic of the plant species under consideration. In fact, it is possible to characterise the different varieties of species by a true phenolic fingerprint which can be used for the certification of new varieties obtained by hybridisation. These same approaches can be used to detect certain frauds in agri-food products (Fleuriet and Macheix, 2003).

Within a single plant species, there are also major differences between the various varieties or cultivars, sometimes greater than between the species themselves. The example of dry beans is already significant in terms of anthocyanins: some varieties (red beans) are very rich in anthocyanins at the time of ripening, while others (white beans) are completely devoid of anthocyanins (Leighton *et al.*, 1992).

5. Polyphenolic compounds in dry bean kernels

Dried vegetables are rich in total phenolic compounds, with levels ranging from 300 mg to 17 g.kg ^{1} (Fleuriet and Macheit, 2003). In the case of dried beans, the colour of the grain is determined by the presence of polyphenolic compounds, located mainly in the hull (11% of the weight of the grain) with low or negligible quantities in the cotyledons (Gonzalez de Mejia *et al*,.1999). The main polyphenolic compounds are flavonoids, such as flavonol glycosides, anthocyanins and condensed tannins (proanthocyanidins), which form the most widely distributed group in beans (Aparicio *et al.*, 2005), mainly in the hull, with a content of 9.4 to 37.8 mg catechin equivalent/g (Beninger and Hosfield, 2003). In general, flavonoid components showed distinct differences. Black beans contain mainly the *3O-glucosides* delphinidin, petunidin and malvidin, whereas pinto beans contain kaempferol and its *3O*-glycoside derivatives. Light red kidney beans contain traces of quercetin *3O-glucoside* and its malonates, but pink and dark red kidney beans contain the diglycosides of quercetin and kaempferol. Kidney beans contain kaempferol *3O-glucoside* and pelargonidin *3O-glucoside*, while no flavonoids were detected in alubia, cranberry, northern, and flageolet beans (Figure 03) (Lin *et al.*, 2008).

According to the work of Luthria and Pastor-Corrales (2006); Granito *et al.* (2007), the total phenolic acid content extracted by HPLC was 31.2 mg/100 g, of which ferulic acid was the most abundant phenolic acid, followed by *p-coumaric* acid and sinapic acid, which were extracted from the bean samples analysed.

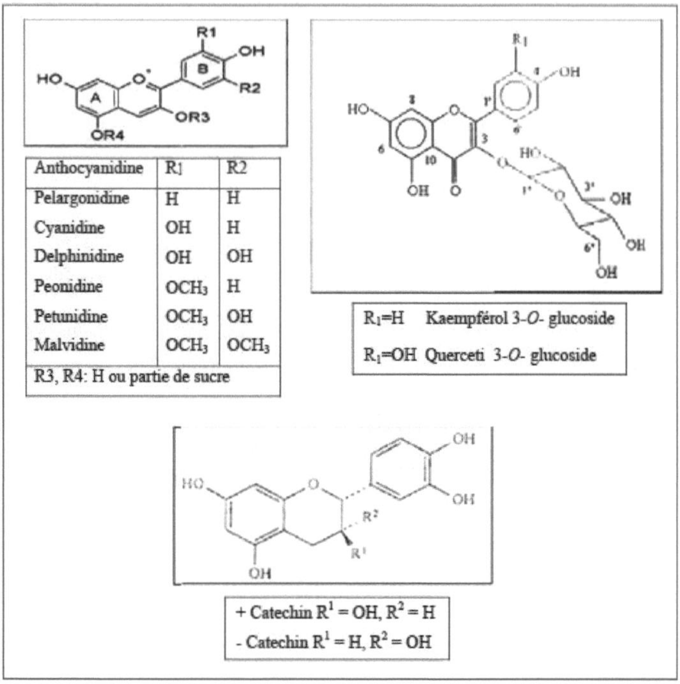

Anthocyanidin	R1	R2
Pelargonidine	H	H
Cyanidine	OH	H
Delphinidine	OH	OH
Peonidine	OCH3	H
Petunidine	OCH3	OH
Malvidine	OCH3	OCH3
R3, R4: H or part of sugar		

R1=H Kaempferol *3-O-* glucoside

R1=OH Querceti *3-O-* glucoside

+ Catechin R1 = OH, R^2 = H

- Catechin R1 = H, R2 = OH

Figure 03. Chemical structure of some polyphenolic compounds present in common beans (Reynoso *et al.*, 2006).

6. Extraction, characterisation and assay of phenolic compounds

6. 1. extraction

The extraction process is systematically carried out using a polar organic solvent such as ethanol, acstone or methanol (Ross *et al.*, 2009).

6. 1.1 Extraction of simple phenols

Most simple phSnols can easily be extracted with mShanol/water mixtures (80/20, v/v). As they are easily oxidised, it is recommended to work at a tempërature of 0 to 4°C and to ensure protection by adding a reducing agent (ascorbic acid or sodium mëta bisulphite) to the extraction medium, which also makes it possible to avoid hydrolysis of fragile phënolic hëtërosides during extraction. After ëlimination of the alcohol by ëvaporation in vacuo, it is necessary to purify the overall extract thus obtained, firstly by ëeliminating the chlorophyll and carotenoid pigments (extraction with petroleum ether), then by extracting the phenolic compounds with an intermediate polarity solvent such as ethyl acetate. Most of the phenols are then found in this solvent, which it is then easy to remove under vacuum in order to finally transfer the correctly purified phenolic fraction to the methanol (Macheix *et al.*, 2005).

6. 1. 2. Extraction of anthocyanins and apolar flavonoids

Anthocyanin pigments are extracted using a methanol/HCl mixture. In contrast, certain apolar flavonoids have to be extracted using more apolar solvents, such as chloroform or hexane (Macheix *et al.*, 2005).

6. 1. 3 Extraction of condensed forms

Condensed forms of phenolic compounds also pose extraction problems. While tannins can still be extracted correctly using aceton/water mixtures, this is not the case for phenols insolubilised in the pecto-cellulosic walls and in the cutin or suberin. In such cases, chemical or enzymatic hydrolysis of the cell walls can be used to free the phenolic compounds before analysing and assaying them. In extreme cases such as lignin, extraction is only possible after oxidative degradation (e.g. by hydrogen peroxide), which leads to depolymerisation and progressive solubilisation (Macheix *et al.*, 2005).

6. 2. Overall dosage

There is no method that allows all the phenolic compounds present in a plant extract to be measured satisfactorily and simultaneously. Nevertheless, a rapid estimate of the total phenol content can be obtained by various methods, in particular by using a mixture of phosphomolybdate and phosphotungstate marketed under the name of Folin Ciocalteu reagent (Harborne, 1989). This method is very sensitive, and a good approximation can be obtained of the phenol content of the extract, which is then expressed in relation to a reference compound (e.g. gallic acid) (Macheix *et al.*, 2005).

***6. 3.* Characterisation and determination of phenolic compounds by spectrophotometry**

Spectrophotomëtrie is a sensitive quantitative and qualitative mëthod, and allows ëchantIIIons to be analysed at low concentrations. The spectra are cmracteristic to the molecules, and provide information about the molecular backbone and different substitutions (Madi, 2010).

6. 4. Chromatographic separations on paper, thin films and columns

Prior to the introduction of high performance liquid chromatography, paper, thin layer and column chromatographic preparations had also been developed for the separation of phenolic compounds (Van Sumer, 1989).

6. 4.1 Chromatographic separation on paper

This technique is based on the principle of product partitioning. It consists of a stationary phase, the paper, and a mobile phase, the solvent, which carries the separated molecules along at different speeds and distances, depending on their degree of affinity with the two phases. It is generally used for the separation and purification of products in small quantities (Madi, 2010).

6. 4.2 Thin layer chromatographic separation

This method is based on the separation of the various constituents of an extract according to their migration strength in the mobile phase (Ekoumou, 2003; Debete, 2005), which is generally a mixture of solvents adapted to the type of separation required and their affinity for the stationary phase, which may be a polyamide or silica gel. This allows us to fingerprint the polyphenolic content of the extract (Ferrari, 2002).

6. 4.3 Chromatographic separation on column

This technique is based on the adsorption phenomëne, whose molecules are carried down the column at variable speeds, depending on their affinity for the adsorbent and their solubility in the eluent; apolar products are eluted first (Madi, 2010).

6. 5 Separations by high-performance liquid chromatography and capillary electrophoresis

6. 5. 1 Separation by high-performance liquid chromatography

High-performance liquid chromatography (HPLC) is by far the most efficient and widely used technique for the separation and determination of phenolic compounds (Wulf and Nagel, 1978; Moller and Herrmann, 1982). It requires only a small quantity of plant sample and allows the qualitative and quantitative analysis of a complex phenolic extract to be combined in a single, rapid and reproducible operation, which has made it possible to study a wide variety of plant materials (Macheix *et al.*, 1990; Pietta *et al.*, 2003).

6. 5. 2 Separation by capillary electrophoresis

Alongside HPLC, another high-performance separation technique for phenolic compounds is capillary electrophoresis (CE), which was developed for this type of molecule and whose sensitivity is approximately 10 times greater than that of HPLC. It leads to separations different from those obtained by HPLC and has good applications in the separation of the many flavonoid glycosides (Markham and Bloor, 1998). In addition to the quality of the separations obtained, one of the major advantages of separating polyphenols by HPLC or EC is the possible automation of the analyses, which can be combined with UV detection and assay of each of the compounds and with mass spectrometry, which provides information on their chemical structure (Swinny and Markham, 2003).

6.6 Determination of the structure of phenolic compounds by physical-chemical methods

Various types of hydrolysis have long provided valuable information on the chemical bonds between the different molecules forming a native phenolic compound, in particular between the aglycone and the carbohydrate part (Markham and Bloor, 1998). However, these approaches are highly inadequate for elucidating the fine structure of these compounds, which is why the physico-chemical techniques traditionally used by chemists (mass spectrometry, nuclear magnetic resonance, Raman spectrometry, etc.) are used (Macheix *et al.*, 2005).

Mass spectrometry can be used to prove the identity of phenolic compounds previously separated by HPLC or EC by providing information on their molecular mass and the main chemical groups present, thanks to the fragmentation of the molecule (Kuhnie, 2003).

The nuclear magnetic resonance (NMR) technique is a powerful tool for the determination of new, unknown phenolic structures, and numerous examples from various classes, particularly the flavonoids, have been described (Swinny and Markham, 2003).

Chapter 04

Materials and methods

1. Plant material

1.1. Presentation of dry bean varieties

This study focused on two dry bean variëtës. The first variëtë *"MGT djedida"* is a local variety characterised by a dark red colour, medium size and smooth texture. The second variety *"Tima"* is an imported variety, characterized by a white colour, small size and bulbous (Appendix 01). The local variety *"MGTdjedida" was* supplied to us by the C.C.L.S of El Khroub, wilaya of Constantine, it was harvested in July 2010 and was stored in paper bags at environmental temperature. The *'Tima'* variety was imported from France and purchased from a sales outlet for dried vegetables intended for human consumption in the wilaya of Jijel. It was stored in 25kg perforated plastic bags at environmental temperature. The choice of these two varieties was based on the difference in grain colour, origin and genetics.

1. 2. Grain sorting

For each variety of dry bean, a preliminary separation into two batches was carried out. One batch of beans assumed to be healthy (GS) and another for beans assumed to be contaminated (GC). This separation was carried out on the basis of a certain number of morphological criteria such as :

- ❖ a change in the colour of the tegument (outer envelope);
- ❖ a change in the appearance of the grain (grain wrinkle);
- ❖ the change in the shape of the grain;
- ❖ the presence of scars or wounds on the grain, etc. (Botton *et al.*, 1990).

2. Methods

To avoid biased results, all the tests were repeated three times.

2.1. Measuring the moisture content of dry beans

The moisture content of flour from whole dry beans is determined by oven drying at a temperature of 110°C of a 05g test sample to a constant weight and is expressed as a percentage (Francois, 2004):

$$H (\%) = (M1 - M2) / M1 \times 100$$

With :

H (%): Moisture content expressed as a percentage.

M1: Weight of sample in grams before drying.

M2: Sample weight in grams after drying.

2.2 Searching for and identifying moulds

2. 2. 1 Isolation

For mould isolation, we used the ulster or direct method. This is the method preferred by several authors, notably Nguyen (2007), for detecting and isolating mycetes from grains. With this method, mould development is stimulated by incubating the grains directly on a Potato Dextrose Agar (PDA) culture medium (Appendix 02) (Botton *et al.*, 1990).

In order to determine deep fungal contamination and surface contamination of the grains, after classifying the grains into two batches, 'healthy and contaminated', we disinfected the surface of a sample from each batch with 95% ethanol (Guiraud, 2003).

In petri dishes containing the PDA medium, 08 grains from each batch (GS and GC disinfected and not surface disinfected from both varieties) are placed per dish so that they are sufficiently spaced. These boxes are hermetically sealed with parafilm and incubated at 30°C for three to four days in order to encourage the growth of contaminating moulds in the grains (Dehimat, 1990).

2. 2.2 Purification

Moulds are purified by removing a terminal hypha following culture in a petri dish on fresh medium (PDA). The strain is inoculated into the centre of the dish. The sample is taken when the strain has developed sufficiently (Guiraud, 2003).

3. 2. 3. Identification

It is relatively easy to identify the genus, but it is much more difficult to determine the species with certainty (Guiraud, 2003). The identification of genera has been based on cultural (macroscopic) and morphological (microscopic) characteristics (Botton *et al.*, 1990).

2. 2.3.1 Study of crop characteristics

Cultural characteristics were essentially studied on PDA conic medium in pëlr1 boxes and inoculated by touch. Macroscopic examination consists of binocular or naked eye observation of the cultural characteristics of the colonies (Smith, 2002).

The cultural characteristics thus studied are :

- growth rate ;
- the colour of the colonies ;
- the texture of the thallus ;
- the presence or absence of exudations ;
- the colour and the change in colour of the environment.

2. 2. 3. 2 Study of microscopic characteristics

In situ observation of the morphology of the spore chains and that of the mycëlium was carried out using the mounting technique without staining. Microscopic preparations are carried out in the fresh state in a liquid medium between slide and coverslip. The manipulation consists of placing a small mycëlial fragment on a clean slide between two Bunsen burners in the presence of a drop of mounting liquid (AMANN's Lactophënol) (Appendix 02) and carefully dilacëating it with two needles to avoid the realisation of a too dense and unobservable separation (Botton *et al.*, 1990), then cover it carefully with a slide, taking care not to create air bubbles or overflow.

2.2.4 Conservation of strains

Isolated strains are stored in tubes on inclined enclosures (PDA). After subculturing, the cultures are kept for a week at 30°C and then stored at 4°C to promote viability and limit the possibility of variation (Bouchet *et al.*, 1999).

***2.3* Assessment of total polyphenol content in grains**

2. 3. 1 Sampling

The whole beans of the two dry bean varieties are ground using an automatic grinder to obtain flour and better extract the polyphënols.

2. 3. 2 Extraction of total polyphenols

The extraction of total polyphënols from dry bean variëtës (white variëtë: ëchantIШon healthy and ëchantIШon contanine, red variëtë: ëchantIШon healthy and ëchantIШon contanine) is carried out according to the protocol proposed by Luthria and Pastor-Corrales (2006) and modified by Mujica *et al.* (2009).

1g of flour from each sample is solubilised in 25ml of methanol 80% (methanol-distilled water

80:20; v/v) acidified with 0.1% 2N HCl. The mixture is left for 2h at room temperature, then centrifuged at 1800g for 15min. The residue is rë-extracted with 25ml of 80% methanol and centrifuged again. At the end, the supernatants are combined and the dry extract is recovered after dry evaporation (Figure 04).

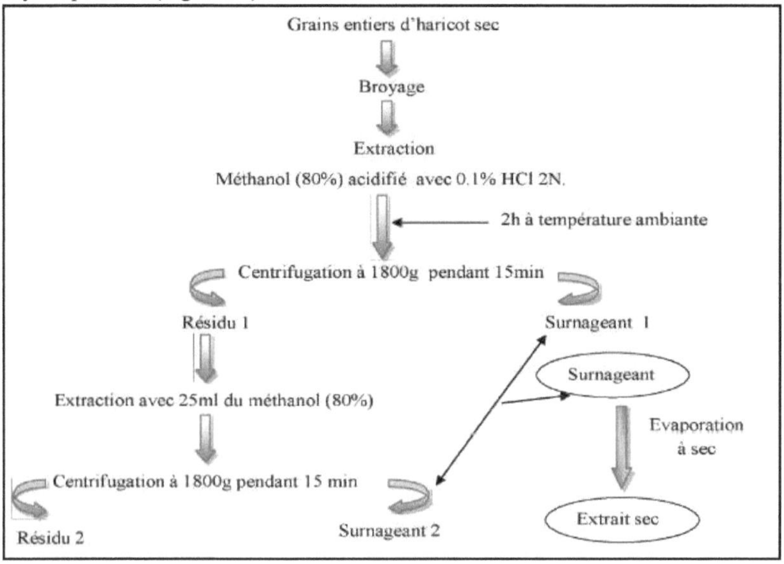

Figure 04. Rëcapitulation of the ëextraction steps for total polyphënols.

2. 3.3 Determination of total polyphenols

The quantification of total polyphënols is based on the Folin-Ciocalteu reaction. This method is chosen for the following reasons:

❖ It is a well-standardised method and meets the criteria of feasibility and reproducibility;

❖ the availability of Folin reagent;

❖ the chromophore's long absorption wavelength (760nm) minimises interference with the sample matrix, which is often coloured (Huang *et al.*, 2005).

2. 3. 3. 1 Principle

All the phenolic compounds are oxidised by the Folin-Ciocalteu reagent. The reagent consists of a mixture of phosphotungstic acid ($H_3PW_{12}O_{40}$) and phosphomolybdic acid ($H_3PMo_{12}O_{40}$), which is reduced on oxidation of the phenols to a mixture of blue oxides of tungsten (W_8O_{23}) and molybdenum (Mo_8O_{23}) (Ribereau-Gayon, 1968).

Absorption is proportional to the quantity of polyphënols present in the vëgëtal extracts (Charpentier and Boizot, 2006).

2. 3. 3. 2. Protocol

According to Singleton *et al* (1999), the assay protocol for total polyphënols consists of adding 250 pl of Folin's reagent diluted with distilled water (50% v/v) to 100 pl of polyphënolic extracts obtained by dissolving 0.02 g of dry extracts in 1 ml of mëthanol. After 5 min incubation at 25°C, 250 pl of 20% (w/v) sodium carbonate (Na2 CO_3) are added to the tubes and the mixture is made up to 2000 pl with distilled water. The absorbance is read at 760 nm after 60 min. The blank is prepared for each sample by filling our polyphenolic extracts with 80% methanol (Figure 05).

The concentration of total polyphenols is calculated from the regression equation of the calibration curve established with gallic acid (0.03-0.50 mg/ml) on the basis of preliminary tests, and is expressed in mg gallic acid equivalent per gram of extract (mg GAE/g extract):

$$T = c \cdot v / m$$

With :

T: total polyphenol content (mg gallic acid equivalent/g grain extract). c: gallic acid equivalent concentration (mg/ml).

v: volume of extract (ml).

m: mass of grain extract (g) (Madi, 2010).

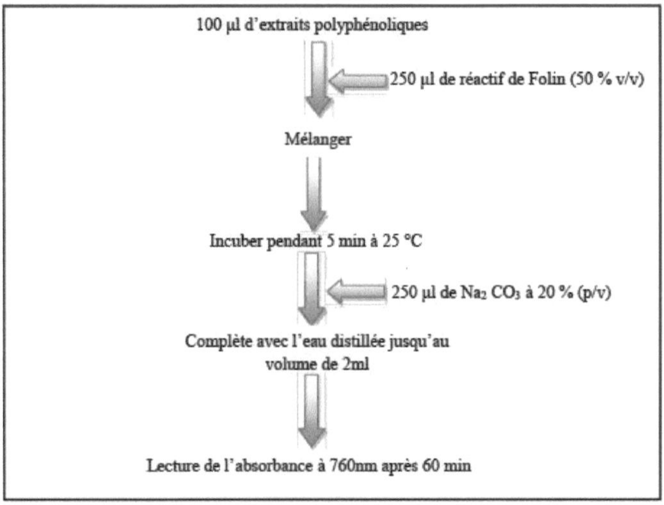

Figure 05. Summary of the total polyphenol assay steps.

2. 4. Testing the antifungal power of polyphenolic extracts

3. 4.1 Choice of solvent for recovering polyphenolic extracts

If polyphënolic extracts are to be subjected to bioassays, the toxicity of the recovery solvent may be open to criticism, as even trace amounts of the solvent should not interfere with the biological process (Yrjonen, 2004). For this reason, we have chosen dimethyl sulphoxide (DMSO), which is the preferred solvent by the majority of authors, including Alavi *et al.* (2005), Mohammedi (2006) and Ownagh *et al.* (2010), who have shown that DMSO has no potent antifungal properties.

4. 4.2 Antifungal test

The antifungal activity of the polyphenolic extracts was tested on six strains. One strain, chosen at random from each genus, was tested *in vitro* by the direct contact method, on gelose "PDA" medium to determine the inhibition rates, by comparing their action at various concentrations on mycelial growth (Hussin *et al.*, 2009) and to estimate the IA_{100}

(l'antifungal index 100), and on liquid medium, Potato Dextrose Broth "PDB" to determine the minimum inhibitory concentration (MIC), fungicidal concentration (FC) and fungistatic concentration (FCS) (Derwich *et al.*, 2010).

5. 4. 2. 1 Direct contact method

a) Determination of the inhibition rate

***Preparation of dilutions of polyphenolic extracts from dried beans**

Based on previous tests, a range of solutions with concentrations ranging from 18.75mg/ml to 300 mg/ml has ële prepared as follows:

-3g of each polyphënolic dry extract is introduced into a tube containing 10 ml of DMSO (300mg/ml);

-5ml of the solubilisë extract is then introduced into another tube containing 05 ml of DMSO (150 mg/ml);

-Proceed in the same way to obtain 75 - 37.5 and 18.75 mg/ml (Figure 06).

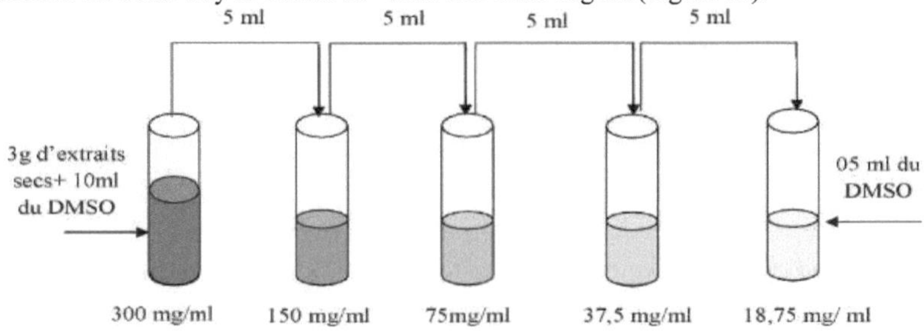

Figure 06. Preparation of dilutions of polyphënolic extracts.

***Seeding**

One ml of each polyphënolic extract, for each concentration, is addedë to tubes containing 19 ml of the still liquid sterile PDA medium. The mixture is homogenised and brought to 45°C (Subrahmanyam *et al.*, 2001). It is then immëdiately poured into 90 mm p'tri boxes (20ml/box) (Satish *et al.*, 2010). After solidification of g'lose, the pë1H dishes are partitioned into six parts (corresponds to the number of strains to be tested) and inoculated with a тycëH^ disc, 6 mm diamëtre taken from the young mycëte culture. PDA without extract was used as a tëmoin for each strain (Mishra and Dubey, 1994; Khallil, 2001).

The final concentrations of polyphënolic extracts used were calculated from the following equation:

$$C_f = C_i/20$$

With :

C_f : final concentration of 1 polyphënolic extract in 1ml of PDA;

c_i: initial concentration of the polyphënolic extract solubilisë in DMSO (Mohammedi, 2006).

***Incubation**

The strains are incubated for 2 days for *Rhizopus sp.*, 4 days for *Alternaria sp.* and *Moniliella sp.,* 7 days for *Aspergillus sp.*, *Penicillium sp.* and *Fusarium sp.,* at a tempërature of 30°C (Mohammedi, 2006).

***Expression of results**

The percentage of growth inhibition тycëHenпe, compared to the tëmoin, a ëlë calcute by the following formula:

$$PI (\%) = (A - B)/ A \times 100$$

Or :

PI(%): Inhibition rate expressed as a percentage;

A: Diamëtre of colonies in "tëmest positive" plates;

B: Diamëtre of colonies in dishes containing 1 grain extract (Bajpai *et al.*, 2010).

The polyphënolic extract is said:

❖ Very active when it has an inhibition of between 75 and 100%; the fungal strain is said to be very sensitive;

❖ Active when it possëde an inhibition of between 50 and 74%; the fungal strain is said to be sensitive;

❖ Moderately active when it has an inhibition of between 25 and 49%; the strain is said to be limited;

❖ Not very active or not active when it has an inhibition of between 0 and 24%; the strain is said to be not very sensitive or resistant (Alcamo, 1984).

b) Determination of the antifungal index ($_{IA100}$)

The concentration that 100% inhibits growth тycëHenпe is expптëe by the AI_{100} . The AI values$_{100}$ have been calculated graphically, where, 1 abscissa is represented by the concentration of 1 polyphënolic extract and i'ordon^e by the percentage of mould growth inhibition (Chang *et al.,* 2008).

2. 4. 2. 2 Liquid dilution method

This technique involves two steps, the first of which determines the minimum inhibitory concentrations (MICs) and the second determines the fungicidal (CF) and fungistatic (CFS) concentrations.

a) Determination of minimum inhibitory concentrations (MICs)

The PDB medium was first prepared and sterilised, then stored in 250 ml vials.

***Preparation of fungal strain suspensions**

After sporulation of the selected fungal strains (sown on 20 ml of PDA in pël:rl dishes), the spores of the young cultures are rëcupërëes by addition of 10 ml of sterile distilled water under vigorous agitation (Solis-Pereira *et al.*, 1993). Afterwards, the absorbance (reading at a wavelength of 625 nm) of the fungal suspension is ëevaluated; this evaluation is carried out with the aim of standardising the spore suspension at 10^6 spores/ml (Hossain *et al.*, 2008). It is estimated that an absorbance of between 0.08 and 0.1 corresponds to a concentration of 10^6 spores/ml (Braga *et al.,* 2007).

***Preparation of dilutions of polyphenolic extracts**

The polyphenolic extracts of the grains of the healthy white variety (PHBS), the contaminated white variety (PHBC), the healthy red variety (PHRS) and the contaminated red variety (PHRC), solubilised in DMSO, are added to the PDB at a rate of 1 ml in 9 ml to give a concentration of 30 mg/ml. The PDB is then diluted successively to give dilutions of 15 - 7.5 - 3.75 - 1.875 mg/ml. These concentrations have ële chosen after preliminaries tests.

***Seeding**

A 10 pl spore suspension of the fungal strains to be tested was inoculated into test tubes containing PDB medium at different concentrations; these tubes were incubated for 2 to 7 days at 30°C. In parallel, a tube containing the PDB medium was inoculated only with the fungal spore suspension; it will serve as a control.

***Reading and expressing results**

Minimum concentrations for which no growth has been observed are defined as minimum inhibitory concentrations (Bajpai *et al.*, 2008).

b) Determination of fungicidal (CF) and fungistatic (CFS) concentrations

For tubes in which no growth is observed, the experiment is continued in petri dishes. Each dish containing 20ml of stërile PDA is inoculated with I0ul of each tube showing total inhibition of fungal growth. Growth was monitored for 1 to 4 days at a temperature of 30°C. When there is no resumption of growth, the concentrations are said to be fungicidal (CFs) (Zarrin *et al.,* 2010) and the concentrations for which there is growth are said to be fungistatic (CFSs) (Bajpai et *al.,* 2010).

2. 5. Statistical processing

The averages plus or minus the standard deviations of the trials and the graphical representations were produced using Excel 2007.

The results obtained were processed by two independent analyses using XLSTAT (2008) software. The first analysis is an analysis of variance (ANOVA) to determine the significance of the differences (the significance threshold is <0.05). The second analysis consists in showing correlations between the variables studied (moisture content, contamination rate and total polyphenol content) at the significance level <0.05.

Chapter 05

Results and discussion

1. Moisture content of dry bean samples

The rates cПшпшШе revel es vary between 11.47% and 12.50%. They are equivalent to 11.47±0.15%, 12.50±0.05%, 11.60±0.05% and 12.00±0.30% respectively for the healthy red variety (HRS), the contaminated red variety (HRC), the healthy white variety (HBS) and the contaminated white variety (HBC) (Figure 07).

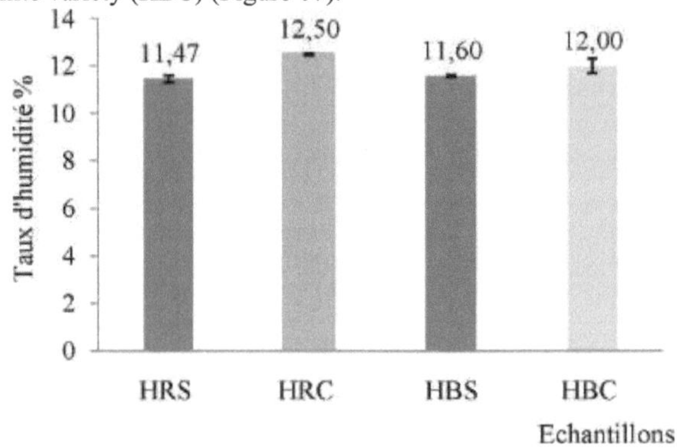

Figure 07. Moisture content of dry bean samples.

Generally speaking, the analysis of variance (ANOVA) indicated a significant difference ($p<0.05$) between the two varieties (white and red). In terms of sampling, no significant difference ($p<0.05$) was found between the healthy batch and the contaminated batch for the white variety, but it was significant for the red variety (Table 08).

Table 08: Difference analysis using Tukey's test between dry bean samples for moisture content ($p < 0.05$).

Samples	Estimated average		
	Healthy sample (HS)	Contaminated sample (HC)	Variety average
Red variety	$11,47 \pm 0,15^{c}$	$12,50 \pm 0,05^{a}$	11,98±0,10
White variety	$11,60 \pm 0,05^{b}$	$12,00 \pm 0,30^{b}$	11,80±0,17

The same letter means no significant difference

Dried beans are foodstuffs likely to be contaminated by moulds before harvest, in the field, during drying or during storage. While numerous studies are available on mycotoxic contamination of grains, investigations into fungal contamination of these raw materials are rarer. In a дёпёгак way, they ёvidence a relationship between the fungal flora and the huniidk rate of grains during storage (Tabuc, 2007).

Moulds are all the more formidable because they tolerate very low levels of cПшпшШё. A huniidity level above the critical value dёclenches fungal aheration of the grains. According to Pitt and Miscamble (1995), a 10-15% dulin content favours the development of storage moulds, the most common of which are *Aspergillus* and *Penicillium*. According to Multon (1982), the degree ofiniidity required during grain storage should not exceed 11% to avoid such alteration. According to the results obtained, the dry bean samples analysed contained sufficient water to cause mould growth. However, these results seem to be different from those found by Laib (2009), where the moisture levels гёуёкз ёwere greater than 15%. This difference seems quite normal to us, since grain moisture depends on several factors including variety, geography, harvest period, storage temperature, etc.

We would like to point out that for our varieties, apart from their origins, a number of factors are ignored, in particular harvesting conditions, storage conditions and storage time. For example, attacks by various pathogenic agents, including moulds, are increased if the grains are stored for a long time (Cruz and Diop, 1989).

2. Isolation and enumeration of moulds

Of the two dry bean varieties, 43 strains were revealed by the direct method and are distributed as follows:

- ❖ Three strains from the superficially disinfected sample of the healthy red variety. These strains are coded: HRSD1, HRSD2, HRSD3.
- ❖ Five strains from the uninfected sample of the healthy red variety. These strains are coded: HRS1, HRS2, HRS3, HRS4, HRS5.
- ❖ Five strains from the superficially disinfected sample of the contaminated red variety. These strains are coded: HRCD1, HRCD2, HRCD3, HRCD4, HRCD5.
- ❖ Seven strains from the non-disinfected sample of the contaminated red variëtë. These strains are codëed: HRC1, HRC2, HRC3, HRC4, HRC5, HRC6, HRC7.
- ❖ Three strains from the superficially disinfected sample of the healthy white variety. These strains are coded: HBSD1, HBSD2.HBSD3.
- ❖ Seven strains from the non-disinfected sample of the healthy white variety. These strains are coded: HBS1, HBS2, HBS3, HBS4, HBS5, HBS6, HBS7.
- ❖ Three strains from the superficially disinfected sample of the contaminated white variety. These strains are coded: HBCD1, HBCD2, HBCD3.
- ❖ Ten strains from the non-disinfected sample of the contaminated white variety. These strains are coded: HBC1, HBC2, HBC3, HBC4, HBC5, HBC6, HBC7, HBC8, HBC9, HBC10.

3. Contamination rate of dry bean samples

3. 1. Contamination rate of the red variety

Beans of the red dry bean variety were found to be contaminated with 20 of the 43 strains isolated, corresponding to a contamination rate of 46.51% (Table 09).

Table 09: Contamination rate of the red variety of dry bean.

Samples	Codes	Contamination rate (%)	Total (%)
healthy disinfected	HRSD	6,97	18,60
sound, not disinfected	HRS	11,63	
contaminates disinfects	HRCD	11,63	27,91
contaminated not disinfected	HRC	16,28	

a) Healthy red variety

Despite the healthy appearance of the grains of the red variety analysed, their contamination rate was relatively high (18.60%). The density of fungal strains in the healthy, non-disinfected batch (11.63%) was greater than that in the healthy, disinfected batch (6.97%), meaning that 4.66% of the contamination was truly superficial and 6.97% was deep contamination.

b) **Contaminated red variety**

The contamination rate of supposedly contaminated dry red beans (27.91%) was more ëкуë than that of supposedly healthy beans (18.60%). Surface disinfection reduced the contamination rate by 4.65%.

3. 2. Contamination rate of the white variety

Grains of the white variëtë were rëyëкз contaminated by 23 strains out of 43 isotees, corresponding to a contamination rate of 53.49% (Table 10).

Table 10. Contamination rates of white variëtë of dry bean.

Samples	Codes	Contamination rate (%)	Total (%)
healthy dësinfectë	HBSD	6,98	23,26
healthy not dësinfectë	HBS	16,28	
contamiiK' dësinfectë	HBCD	6,98	30,23
contamiiK' non dësinfectë	HBC	23,25	

a) **Healthy white variety**

The grains of the supposedly healthy white variëtë were 23.26% contaminated. Surface disinfection of this batch appears to reduce contamination to 6.98%, which means that 9.30% were moulds 1oca1isëes on the surface of the grains.

b) **Contaminated white variety**

Kernels of the white variëtës supposës contamiires turned out to be 30.23% ak'rcs, of which 6.98% of the strains are 1oca1isëes deep.

Moulds are undemanding hëtërotrophic microorganisms, requiring only a supply of carbon and nitrogen for their growth (Abdel Massih, 2007). Dry beans, with their high starch and protein content, are a suitable medium for providing the mould with the necessary nutrients for growth (Kassemi, 2006). The rigid wall of the fungal cell prevents it from phagocytosing the bean's complex nutrients; the mould is obliged to first transform them into simple, absorbable modules. This is made possible by dëpo1ymërases. Under their action, complex po1ymëers such as starch are transformed into simple molecules (Tabuc, 2007), which could explain the high fungal dëtectëe density.

Despite the healthy appearance of the grains, they are rëyë1ëз contaminated. This contamination is due to the initial spore load (Tahani *et al.,* 2008) which, after incubation, find a favourable environment for germination and mycëlium formation. The low fungal density in the seeds could be explained by the fungicidal and fungistatic effect of ethanol (the disinfectant used) (Guiraud, 2003). The latter has no effect on the so-called internal or deep strains of the grains.

The contamination rate of the white variety is higher than that of the red variety. The white variety, exposed to the market without adequate packaging, is exposed to a hot and humid environment, which favours the development of mould (Nguyen, 2007). Although the two varieties do not share the same harvesting and storage conditions, the red variety seems to be more resistant to fungal infestation. In terms of chemical composition, the white variety has the advantage of containing higher levels of protein and carbohydrates than the red variety (Kassemi, 2006). Several authors, notably Abdel Massih (2007), have shown a positive correlation between fungal infestation and grain richness in these two elements. Similarly, the red variety contains more phenolic compounds, hence its colour (Beninger and Hosfield, 2003), which are known to be resistant to pathogenic attack (Cherif *et al.*, 2007). Polyphenolic compounds offer a very appropriate control method in economic terms.

Faced with the damage caused by fungal infestation of dry bean crops, research has encouraged either the search for resistant varieties or the introduction of resistance genes from two donor species *Phaseolus coccineus* L. and *Phaseolus polyanthus* Greenm, phyletically the closest to *Phaseolus vulgaris, which* favours the creation of interspecific crosses, where these two species are used as maternal parents in order to favour the introgression of useful genes from these two taxa (Baudoin *et al.*, 2004).

4. Gender identification

The identification of fungal genera was essentially carried out using the determination keys of Botton *et al.* (1990), Chabasse *et al.* (2002) and Guiraud (2003).

4. 1. Macroscopic study

The main macroscopic characteristics of the different fungal strains isolated from red and white varieties of dry bean are collected in Tables 11, 12, 13 and 14. These tables summarise: the appearance of the тусёНит, the growth, surface area and consistency of the colonies, the colour of the reverse side of the dish, the presence or absence of exudation and pigments characteristic of each strain.

4. 2. Microscopic study

The microscopic study involved the observation of the characteristic structures of 43 isolated fungal strains (Table 15).

The revered strains belong to six mould genera. These genera are : *Alternaria, Aspergillus, Fusarium, Moniliella, Penicillium* and *Rhizopus.* Four of these genera have already been identified and cited by Laib (2009) in dry beans exposed to the local market in the Skikda region. The other two genera, *Fusarium* and *Moniliella*, are found only on the imported variety. This could probably be explained by the poor conditions in which the dried tegumes are exposed to the market, usually without adëquat packaging. This method of sale is practically identical in the eastern Adriatic.

Further ëtudies on the fungal flora of legumes showed that the storage mycetes of beans (*Phaseolus vulgaris*) in India included the genera *Alternaria*, *Aspergillus*, *Cladosporium*, *Colletotrichum*, *Fusarium*, *Penicillium, Rhizoctonia, Stemphylium* and *Trichoderma* (Sud *et al.*, 2005). Species belonging to the genera *Penicillium* and *Eurotium* were common in Taiwanese beans, but in Canadian beans, the most common mycetes were *Alternaria, Fusarium* and *Rhizoctonia* (Tseng *et al.*, 1995).

Table 11. Macroscopic characteristics of isolde strains of healthy red bean varidtd.

Code of strains revealed	Mycelium aerien	Reverse side of the colony	Colony appearance	Surface	Exudation	Pigmentation	Growth	Some photos of the aerial mycelium
HRS1 HRS2	Violet pink	Cream-coloured streak	Downer	Dome-shaped in the centre, then flat and so on	Absence	Absence	Slow and uneven	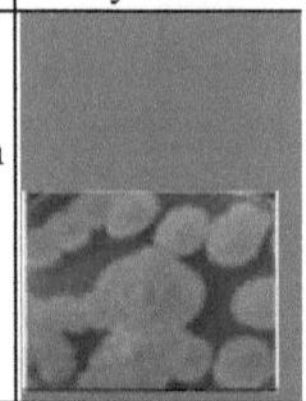
HRS3 HRS4 HRS 5	Olive green with white tips	Green then black, streaked, with light ends	Downer	Flat, domed, then flat	Presence	Absence	Not very fast	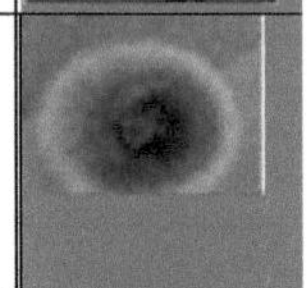
HRSD1 HRSD2	Olive green with white tips	Green then black, streaked, with light ends	Downer	Dome-shaped in the centre, then flat	Presence	Absence	Not very fast	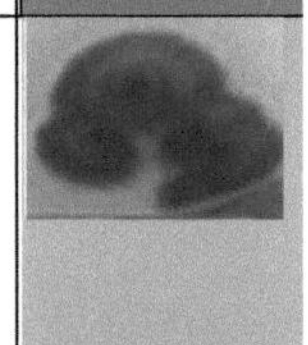
HRSD3	Violet pink with white tips	Cream-coloured streak	Downer	Dome-shaped in the centre, then flat and so on	Presence	Absence	Slow and uneven	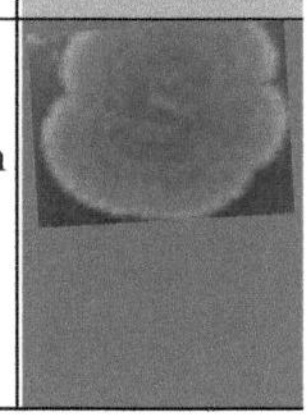

Table 12. Macroscopic characteristics of isolde strains of red varidtd contaminated with dry bean.

Code of strains revealed	Mycelium aerien	Reverse side of the colony	Colony appearance	Surface	Exudation	Pigmentation	Growth	Some photos of the aerial mycelium
HRC1 HRC2	White at first then turns brown	Cream	Cottony and invasive	Sticks to the lid of the box	Absence	Absence	Fast	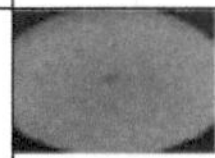
HRC3 HRC4 HRC5 HRCD1 HRCD2	Violet pink with white tips	Cream streak	Downer	Dome-shaped in the centre, then flat and so on	Absence	Absence	Slow, uneven	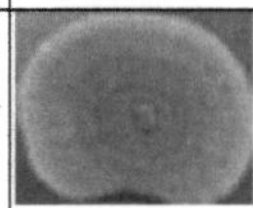
HRC6 HRCD3	Olive green with white tips	Green then black, with light ends	Downer	Plate	Presence	Absence	Not very fast	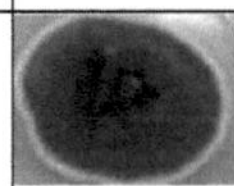
1Y1 HRC7	Olive green with white centre and white ends	Green then black, streaked with a light centre and ends	Downer	Dome-shaped, then flat	Presence	Absence	Not very fast	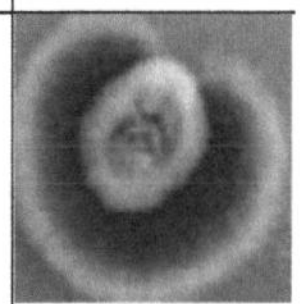
HRCD4	Violet pink with white tips	Cream	Downer	Dome-shaped in the centre, then flat and so on	Presence	Absence	Slow, uneven	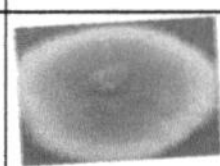
HRCD5	Green	Yellow with green tips	Downer	Dome-shaped in the centre, then flat	Absence	Absence	Slow, uneven	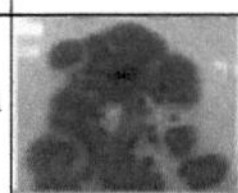

Table 13. Macroscopic characteristics of isolde strains of healthy white dry bean varidtd.

o Code of strains	Mycelium aerien	Reverse side of the	Colony appearanc	Surface	Exudation	Pigmentation	Growth	Some photos of

reveale d		colony	e					the aerial mycelium
HBSD1	Pink with white tips	Cream streak	Downer	Dome shape, then flat and so on	Absence	Absence	Slow and uneven	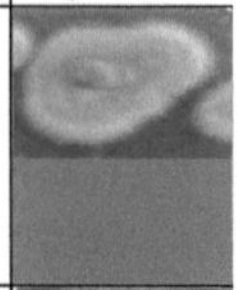
HBSD2	Violet pink	Cream	Downer	Dome shape, then flat and so on	Presence	Absence	Slow and uneven	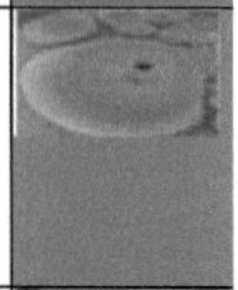
HBS1	White at first, then turns brown	Cream	Cottony and invasive	Sticking to the lid of the petri dish	Absence	Absence	Fast	
HBS2	Brown then turns pale pink	Rose tends towards the matron	Downer	Flat, smooth	Absence	Absence	Slow	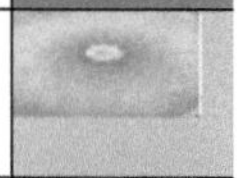
HBS3 HBS4 HBS5	Rose	Cream	Downer	Dome-shaped in the centre, then flat	Absence	Absence	Slow and uneven	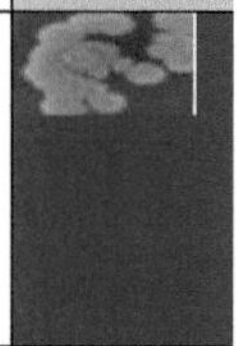
HBS6	White	Light yellow	Creamers	Plate	Absence	Absence	A little slow and homogeneous	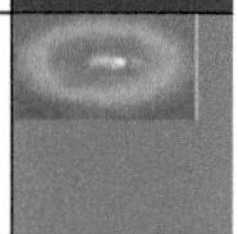
HBS7 HBSD3	White degraded pink	Orange downgraded	Woolworm	Plate	Presence	Presence	Not very fast	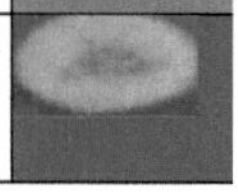

Table 14. Macroscopic characteristics of strains isolated from the contaminated white variety of dry bean.

Code of strains revealed	Mycelium aerien	Reverse side of the colony	Colony appearance	Surface	Exudation	Pigmentation	Growth	Some photos of the aerial mycelium
HBC1	Pink violet aurose grey	Cream streak	Downer	Dome-shaped in the centre, then flat and so on	Presence	Absence	Slow and uneven	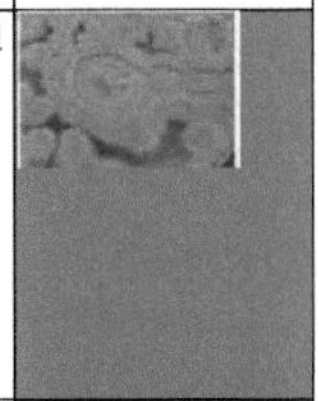
HBC2, HBCD1.	Bottle green turns black	Green-yellow	Powdery, granular	Plate	Absence	Absence	Fast and consistent	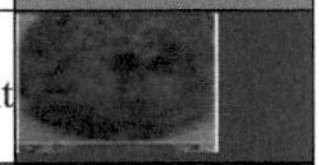
HBC3 o	Fading colour: dark pistachio green in the centre and light at the edges	Light yellow	Powdery, granular	Plate	Absence	Absence	Slow	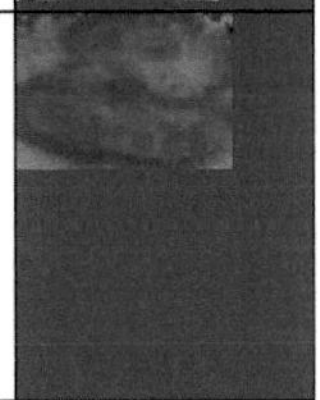
HBC4	Military green	Cream	Downer	Plate	Absence	Absence	Slow and uneven	
HBC5 HBC6 HBC7 HBC8 HBC9 HBC10	Violet pink with white tips	Cream streak	Downer	Dome-shaped in the centre, then flat and so on	Absence	Absence	Slow and uneven	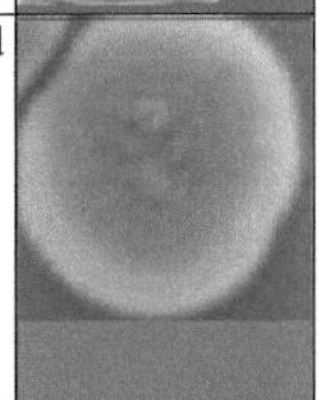
HBCD2 HBCD3	Violet pink with white tips	Cream streak	Downer	Dome-shaped in the centre, then flat	Presence	Absence	Slow and uneven	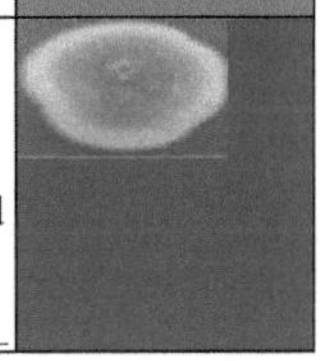

Table 15. Microscopic characteristics of isolde strains of white and red dry bean varidtds.

t:ode des touches rfviJie?	MECHANICAL STUDIES	1 [leulilieali0u dn genre	Quehi пея екепцйе" d^beervwtioD mkroseoplquc (s40)

HBS3, HKS4t HRS5, HRC6, HRC7, HRCDB, i[Rsm. HRS DI	'Conidiophore septc. black, acroissancc sympodiaic;, -Conidia pluriCBIhilaires en cha'ines brunes irregiilieres; SouWilt en fomie de mas sue. c lot sonnets longitudinalcmtntettransvcrsalcmtnt.	*AllenKtriu*	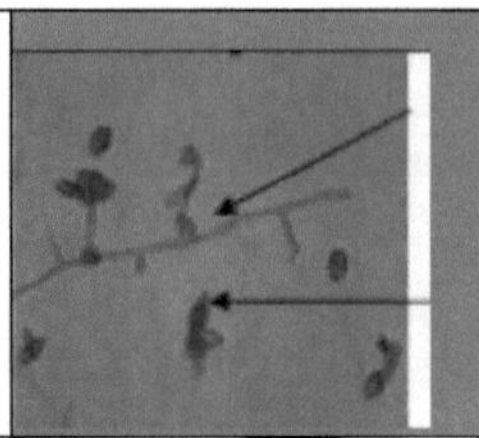	Conidiophore a ero Usance sympod laic ---- Conidic
IIBC2. HBC3, HBCDt	' ihaillc a myedium cloisonne portant de nombreux conidiophores dresses, non rami fids, temiints cnvdsiculcs; -Phiaiidcs farmees dircctement sur la vdsiculeon sur des nittuies; -Teles conidiennes imtScri&ii on bis^rieest 'Masse con idle nnc radiant. conidics cnchaincs nniccl hilaires.	**Я** *sperfttllNS*	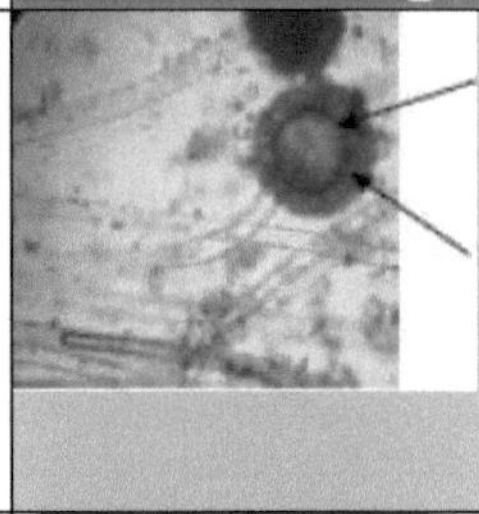	Vdsicnk: x I'hialides and inetuies
HBS6, I[[1S7. HBSD3	-I'hialides plus au moins allongees produisant des mac roton idles courbees. plitrisepiees et poinlnes au\ deUK e\irem lies.	*Ftaarium*	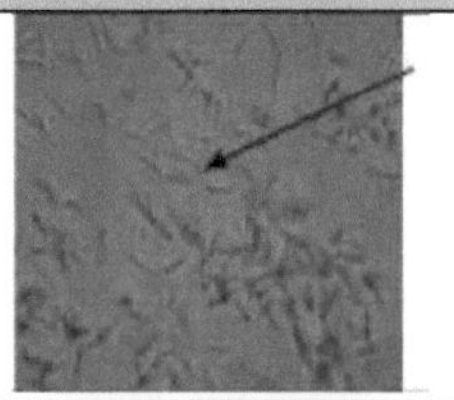	Macroeonidia
HBS2	-Mycdllum cloisonne, fragment in arthospores ; -Hyphae extended by a chain of conidics ; 'Conidics unicllhlaircs cn chape lets rantifidcs sur im conidiophore	*Kiondielia*	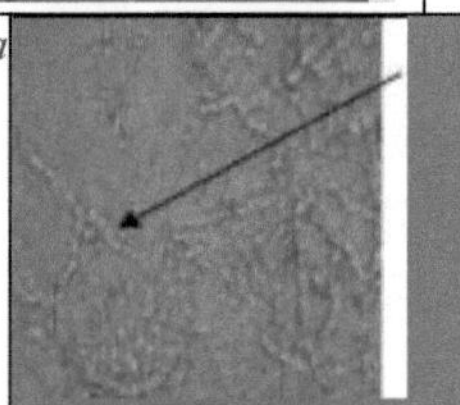	Arthospore
HRS1 HRS2. IIRSIJ3, HRC3, HRC4> HRC5, IIRCDI, HRCD2, IIRCIH HRCD5, IIBSDL 1 HiSI32_, IIBS3 IIBS4. IIBS5 IIBC1. IIRC4, HBC5, IIBC6. HBC7,	Mycelium cloisonne L 'Conidiophores Isolds, ramifids. temiinds par un penicillc j 'Pdnicilllc const ituc de phial ides brane hccs dircctenicnt a Tcxtrcnutc du conidiophore,	*Peniciiil urn*	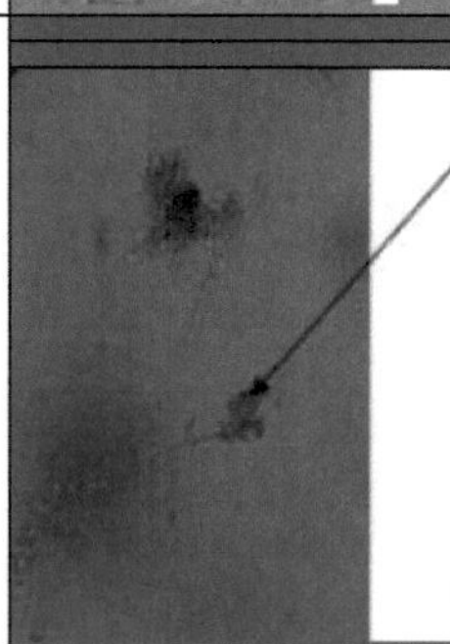	**Pdnicilk**

HBC8, HBC9, IIBC10, HBCD2, HBCD3				
HRC1, IIRC2, HBS1	* [Иal le acroLssance fast ; Sporosystophores generaleinent ires large, term Ines funnel-shaped, Isolds or bouquet tie 2-6 presenting at the base lie rhl/o'ides.	*Rhixopuf*	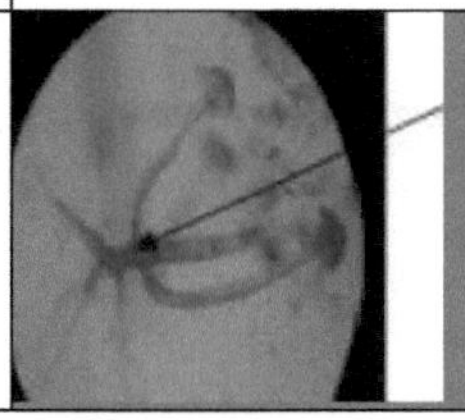	Rh izoKtes

5. Contamination rate of dry bean samples as a function of isolated mould genera

5. 1. Fungal contamination rate of the red dry bean variety

Grains of the red variety of dry bean were contaminated by three types of mould, namely *Alternaria, Penicillium* and *Rhizopus*.

5. 1. 1 Healthy red variety

In the healthy red variety, *Alternaria is the* most dominant genus with a frequency of 62.5%, followed by *Penicillium* with 37.5% (Figure 08).

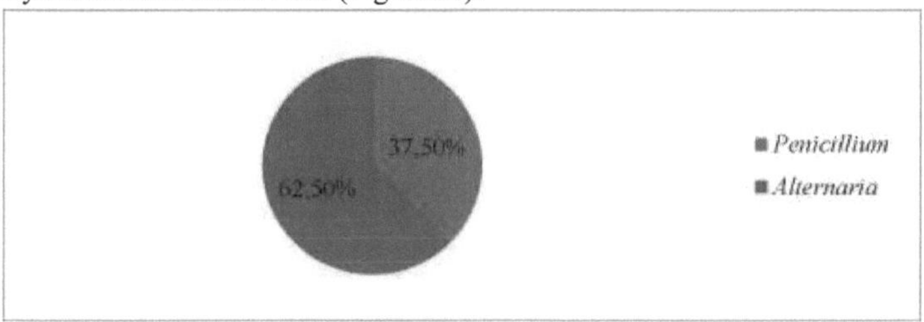

Figure 08. Fungal contamination rate of the healthy red variety of dry bean.

5. 1. 2. Contaminated red variety

For the contaminated red variety, the dominant genus is *Penicillium* (58.33%), followed by *Alternaria* (25%) and finally *Rhizopus* (16.67%) (Figure 09).

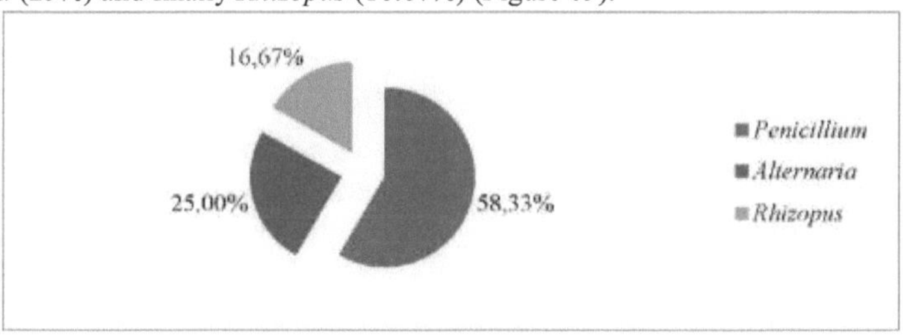

Figure 09. Fungal contamination rate of the contaminated red variety of dry bean.

The dominant genus in the healthy red variety of dry bean is *Alternaria*. Christensen (1994) reports that this genus can be found at harvest; since the red variety is analysed just a few months after harvest, this genus may indeed exist at harvest.

With regard to the genus *Penicillium,* Botton *et al* (1990) have pointed out that, during storage, a flora composed of less cellulolytic and more osmophilic fungi develops. *Penicillium* proliferate mainly on substrates with fairly low humidity (10 to 15%) (Pitt and Miscamble, 1995), which explains the presence of this genus on the dry bean sample with a humidity of 11.47%.

According to Botton *et al* (1990), as the *Penicillium* genus expands, it gradually eliminates fungi that are not very cellolulytic and not osmophilic, belonging mainly to field fungi.

5.2 Fungal contamination rate of the white variety of dry bean

Grains of the white variety of dry bean were found to be contaminated by the genera *Aspergillus, Fusarium, Moniliella, Penicillium* and *Rhizopus.*

5. 2. 1 Healthy white variety

In the healthy white variety, the *Penicillium* genus is the most dominant with a frequency of 50%, followed by the *Fusarium* genus (30%), *Rhizopus* and *Moniliella* with similar frequencies of 10% (Figure 10).

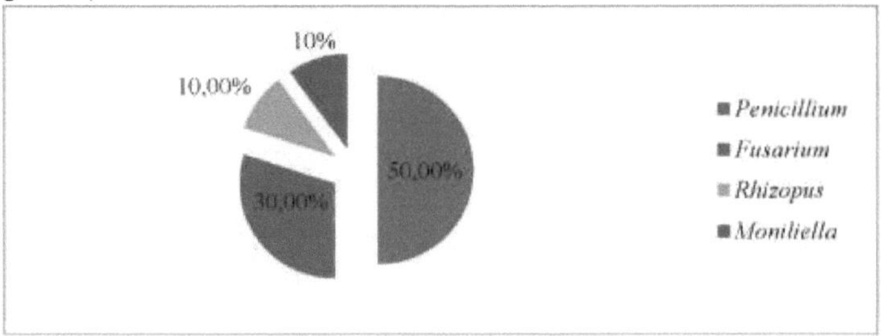

Figure 10. Taux de contamination fongique de la variété blanche saine d'haricot sec.

The genera *Penicillium*, *Fusarium* and *Rhizopus* accounted for 50%, 30% and 10% of the contamination respectively. The genus *Moniliella*, despite the rarity of its isolation (Botton *et al.*, 1990), accounted for 10% of the contamination in the sample analysed.

Referring to Abdel Massih (2007), despite the fact that only humidity values of 22 to 25% cause *Fusarium* development, a non-negligible rate of contamination equivalent to 30% was observed in the sample analysed with a humidity of only 11.60%.

5. *2. 2* Contaminated white varieties

For the contaminated white variëtë, the most dominant genus is *Penicillium* (76.92%), followed by *Aspergillus* with a frequency of 23.08% (Figure 11).

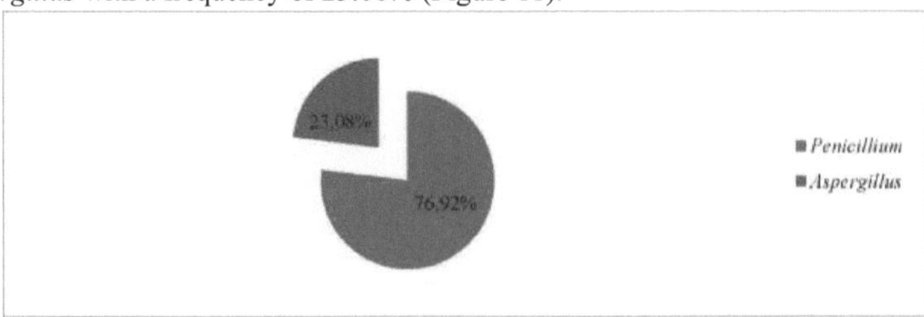

Figure 11. Fungal contamination rate of the contaminated white variety of dry bean.

According to Bourgeois and Larpent (1996), *Aspergillus* and *Penicillium* are the usual hosts of beans, which generally develop during improper storage. Dry white beans sold without

adequate packaging are exposed to the risks of insects and pests, which transmit spores, and to fluctuating temperatures and humidity, factors which favour the multiplication of moulds.
The work of Tahani *et al* (2008) indicates that grains, separated due to a doubt of contamination and the presence of broken grains, show a rapid evolution of their contamination by xerotolerant moulds (*Aspergillus* and *Penicillium).* According to Riba *et al* (2005), the growth of *Aspergillus* and *Penicillium* is favoured by the lack of ventilation coupled with high temperatures. The presence of *Aspergillus* and *Penicillium* in dry white beans suspected of being contaminated may be due to a lack of ventilation after the beans were stacked in stored bags.
The five genera identified in the samples analysed have toxigenic strains. The range of harmful effects of their mycotoxins is very wide, from cancerogenic, mutagenic, necrotizing, neurotoxic, hepatotoxic, hematotoxic, etc. (Brochard and Le Bacle, 2009). Assessment of total polyphenol content
The calibration curve obtained, taking gallic acid at different concentrations as the standard, shown in Figure 12, demonstrates the linear response of the detector as a function of the different concentrations. The choice of this representation model is based on the methodology of several authors, including Mujica *et al* (2009).

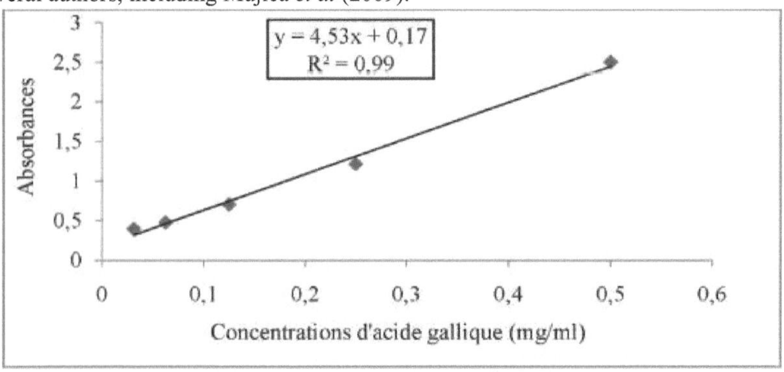

Figure 12. Courbe d'étalonnage d'acide gallique.

Based on the absorbance values of the polyphenolic extracts (Appendix 03), reacted with the Folin-Ciocalteu reagent, and compared with the standard gallic acid solution, the results of the quantitative analysis of the total phenolic compounds are given in Table 16.

Table 16. Total polyphenol content of extracts (mg EAG/g).

Samples	Total phenols (mg EAG/g)	Average (mg EAG/g)
Healthy kidney beans (HRS)	0,42±0,007	0,40±0,005
Contaminated kidney beans (HRC)	0,39±0,003	
Healthy white beans (HBS)	0,32±0,008	0,27±0,005
Contaminated white beans (HBC)	0,22±0,002	

The results recorded above show that there is a varietal effect, with the red variety being richer in total polyphenols (0.40±0.005 mg EAG/g) than the white variety (0.27±0.005 mg EAG/g).

For the red variety, the analysis of variance (ANOVA) showed no significant intra-varietal difference *(p<0.*05%) in the quantity of polyphenols between the healthy batch and the contaminated batch. Conversely, the white variëtë showed a significant intra-varietal difference between the two batches *(*p<0.05%) (Table 17).

Table 17. Difference analysis using Tukey's test between dry bean samples for total polyphenol content *(p <0.*05).

Samples	Estimated average	
	Healthy sample (HS)	Contaminated sample (HC)
Red variety	0,42±0,007[a]	0,39±0,003[a]
White variety	0,32±0,008 [b]	0,22±0,002[C]
The same letter means no significant difference		

As for the varietal difference, the same finding was made by Laparra *et al* (2008); Ying Tan *et al* (2008); Xu and Chang (2009), where coloured dry beans were found to be richer in polyphenols than white dry beans. Bean grain colour is determined by the presence and concentration of flavonol glycosides, anthocyanins and condensed tannins (Reynoso *et al.,* 2006). The red dry bean genotype has a higher anthocyanin content than the white genotype. This difference is probably due to anthocyanins, a well-known group of water-soluble colourants, which contribute significantly to bean colouration (Horbowicz *et al.*, 2008).

Whatever the variety, we can clearly see that healthy samples are richer in total polyphenols than contaminated samples. It seems that the greater the quantity of polyphenols present, the lower the level of contamination. This indicates the existence of a negative correlation between the level of contamination and the total polyphenol content in dried beans. This correlation has been reported by several researchers in the field of plant bioactive substances, including Abad *et al* (2007), who attribute this property to flavonoids, such as flavonol glycosides, the majority class of polyphenolic compounds in dry bean grains.

Referring to the bibliography and with the aim of comparing the total polyphenol levels obtained with those quoted in the literature, several factors must be taken into account which may influence the polyphenolic content. These include climatic and environmental factors (Ebrahimi *et al.,* 2008), genetic factors, experimental factors (Miliauskas *et al.*, 2004), and differences in extraction methods, evaluation and expression of results between authors (Lee *et al.,* 2003). As a result, it was difficult to compare our results with those in the literature. For example, Cardador *et al.*

(2002) found a content of 2.09 mg catechin equivalent/g. In another publication, the same authors reported concentrations of phënolic compounds from 3.28 to 16.61 mg ëequivalent of catëchine/ g in six dry bean cultivars (Oomah *et al.*, 2005). Recently, Mujica *et al.* (2009) found a content of 42.8 to 50.1 mg GAE/g, which is very high compared with our results.

6. Correlations between contamination rate, humidity and total polyphenol content

The main results obtained from the search for possible correlations (Table 18), indicate a positively significant correlation (r = 0.773) between the humidity level and the grain contamination level, which means that the higher the humidity level, the higher the contamination percentage. This correlation can be explained by the fact that most moulds prefer high humidity levels (Moreau, 1996), particularly during the germination phase, which requires a greater input of water than during the growth phase (Basset, 2009). On the other hand, the contamination rate was negatively and significantly correlated (r=-0.698) with the polyphënol content. This inverse correlation can be interpreted in two ways: either the high percentage of contamination leads to a reduction in the total polyphënol content, or it is the presence of a large quantity of total polyphënols that causes a reduction in the grain contamination rate. According to the literature, this correlation is probably due to the fungicidal and fungistatic effects of polyphenols (Esekhiagbe *et al.*, 2009, Kawamura *et al.*, 2010).

Table 18. Correlation matrix (*Pearson (n)*) between the main analyses carried out on the dry bean samples (with a significance level of 0.05).

Variables	Humidity level	Contamination rate	Content in polyphenols
Humidity level	1	0,773*	-0,085
Contamination rate		1	-0,698*
Polyphenol content			1

**: significant correlation*

7. Test of antifungal power

8. 1. Inhibition rate

Inhibition rates (percentage measurements) are representedësentës in Figures 13, 14, 15 and 16 based on the modële of Golam *et al.*(2011). This measurement allowed us to classify the fungal strains according to their degree of sensitivity to each concentration testedëe.

Through the different percentages of inhibition obtained, the polyphënols ëtudiës showed variable activities on the filamentous strains tested. Strains belonging to the *Moniliella* and *Alternaria* genera appeared to be the most sensitive.

For the *Alternaria* strain, polyphënols extracted from contmnine kidney bean (PHRC) and contmnine white bean (PHBC) gave higher inhibition rates than polyphenols from healthy kidney bean (PHRS) and healthy white bean (PHBS). The latter were greater than 50%, for concentrations ranging from 0.94 mg/ml to 15 mg/ml. This can probably be explained by the absence of contamination by this type in the case of contaminated white beans and the reduction in the contamination rate for contaminated red beans if the contamination rate is compared with that of healthy red beans (Figure 13).

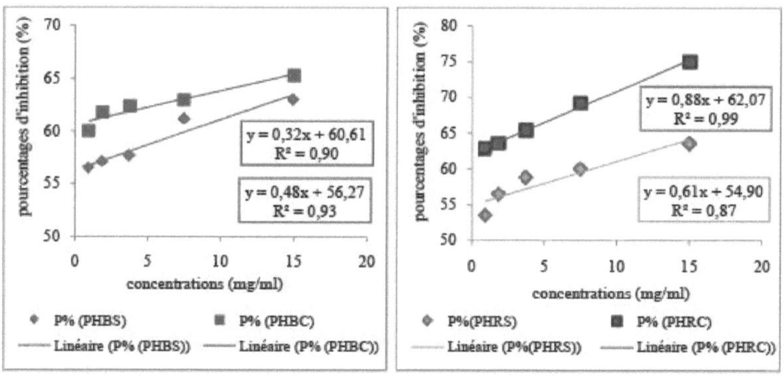

Figure 13. Inhibition rate of polyphenolic extracts against *Alternaria sp.*

As for the strain belonging to the genus *Fusarium,* it showed little sensitivity to polyphenolic extracts of contaminated white beans for all five concentrations (Figure 14). The other types of polyphenolic extracts were moderately active and depigmented the aerial mycelium of the strain in question (Appendix 05).

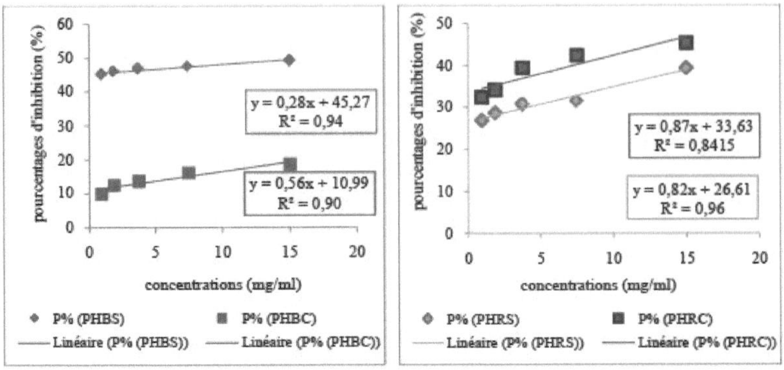

Figure 14. Inhibition rate of polyphenolic extracts against *Fusarium sp*.

For *Moniliella sp.,* three types of polyphenolic extracts were found to be highly active; healthy kidney bean polyphenolic extracts, contaminated kidney bean polyphenolic extracts and contaminated white bean polyphenolic extracts for concentrations ranging from 1.87mg/ml to 15mg/ml (inhibition rate >75%) with the exception of PHRC which was active for concentrations of 0.94 mg/ml and 1.87 mg/ml. Polyphenolic extracts of healthy white beans (PHBS) were found to be active (inhibition rate >60%) at all five concentrations. This justifies the absence of this genus in our samples, with the exception of the low or even negligible contamination rate for healthy white bean grains (a genus rarely isolated) (Figure 15).

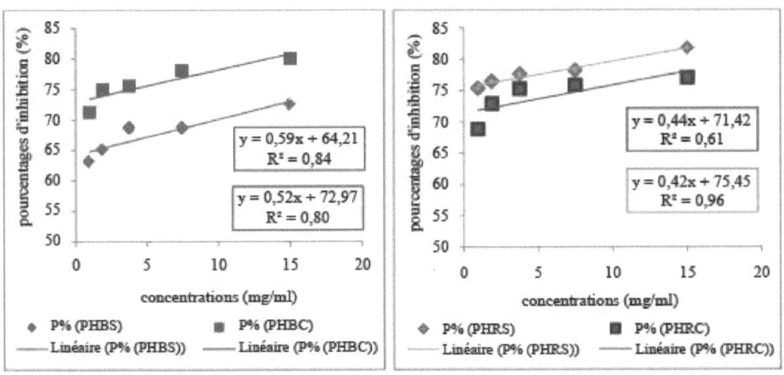

Figure 15. Inhibition rates of polyphenolic extracts against *Moniliella sp.*

For a concentration of 15 mg/ml, the polyphënolic extracts of the 6 healthy samples of the two dry bean variëtës showed inhibition rates above 50% on the genus *Rhizopus, which was* characterised by a decrease in sporulation (Appendix 07). For the other concentrations, this genus was rëyë!ë limile and the polyphënolic extracts are said to be moderately active (Figure 16).

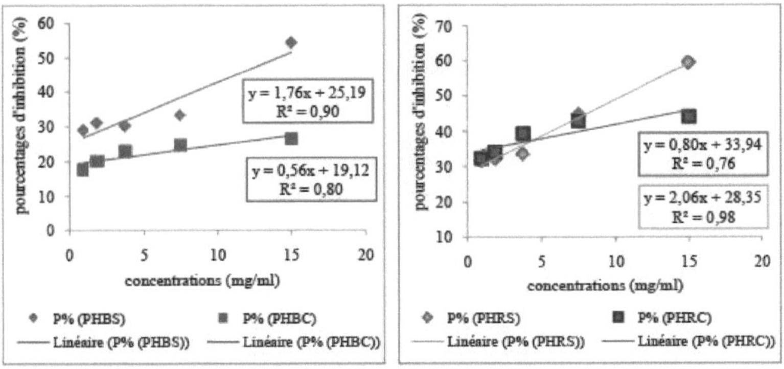

Figure 16. Inhibition rate of polyphënolic extracts against the *Rhizopus sp.* strain.

The *Penicillium sp.* and *Aspergillus sp.* strains grew at a constant rate, which prevented us from calculating their inhibition rates. The results obtained are shown in Figures 17 and 18, where we can clearly see a change in colour.

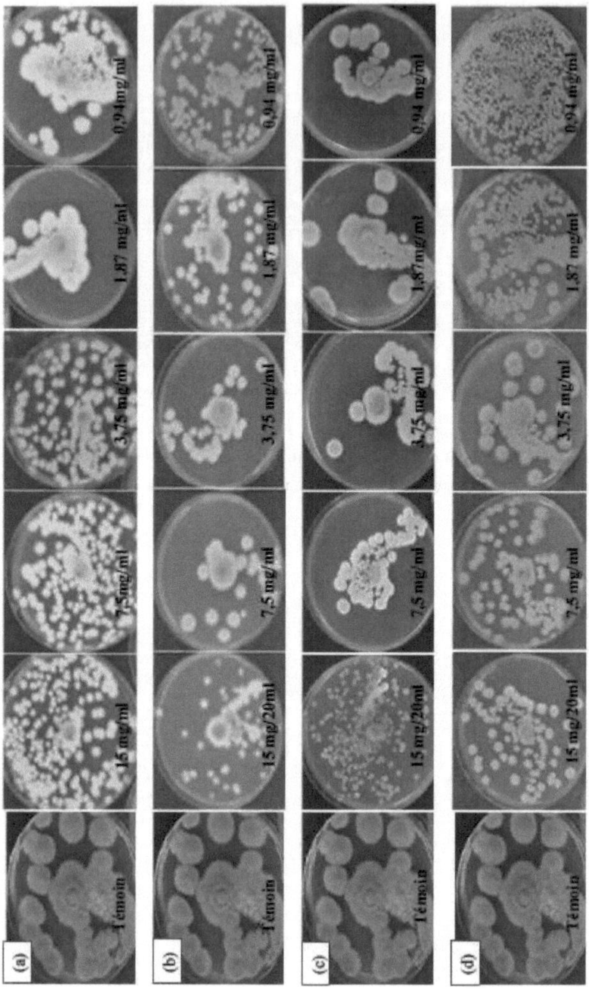

Figure 17. Inhibition effect of polyphdnolic extracts on the *Pemcillium sp*. strain (a): PHRS, (b): PHRC, (c): PHBS, (d): PHBC.

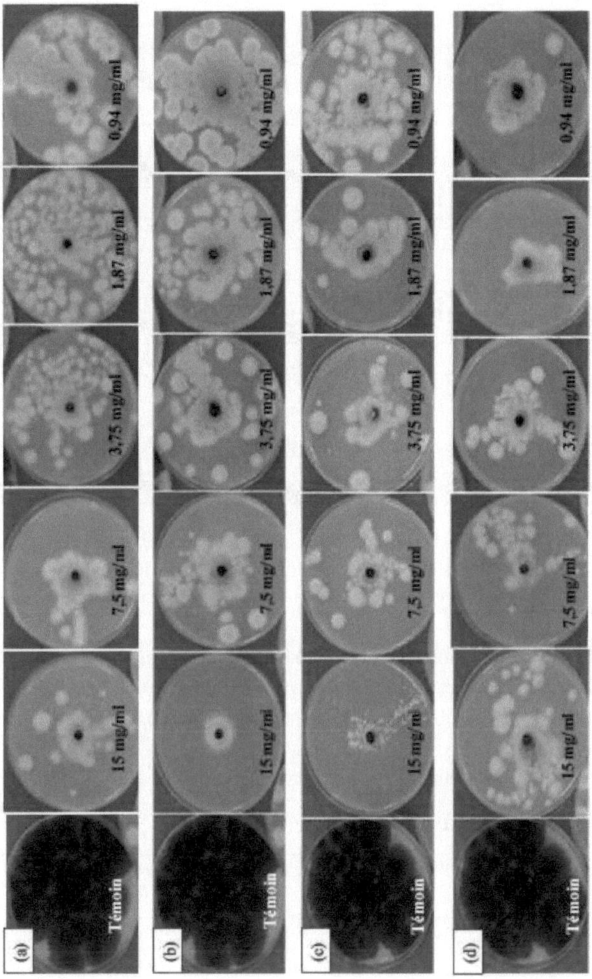

Figure 18. Inhibition effect of polyphdnolic extracts against *Aspergillus sp.* strain (a): PHRS, (b): PHRC, (c): PHBS, (d): PHBC

8. 2 Determination of antifungal indexes (IA100)

The IA100 determined graphically are summarised in Table 19.

Tableau 19. Rĕcapitulation of antifungal indexes (IA_{100}) of polyphĕnolic extracts of ĕcHaтШопз dry beans.

Strain	IA100 (mg/ml)			
	PHRS	**PHRC**	**PHBS**	**PHBC**
Alternaria sp.	73,93	43,10	91,10	123,09
Aspergillus sp.	ND	ND	ND	ND
Fusarium sp	89,5	76,29	195,46	158,95
Moniliella sp	58,45	64,95	60,66	51,98
Penicillium sp.	ND	ND	ND	ND
Rhizopus sp.	34,78	82,57	42,50	144,43

ND: Non Determine

According to the results obtained, the strains tested do not have the same AI_{100} . The AI_{100} of strains belonging to the genera *Aspergillus* and *Penicillium* were not determined due to spore dispersal, so we were unable to measure their diameters.

Despite the existence of relatively low inhibition rates in certain strains, the results obtained prove the existence of antifungal activity against the strains tested. Done, it is interesting to estimate the minimum inhibitory concentrations (MICs), fungicidal concentrations (CF) and fungistatic concentrations (CFS).

8. 3. Minimum inhibitory concentrations (MICs)

Minimum inhibitory concentrations (MICs) must be calculated in order to determine the antifungal efficacy of polyphĕnolic extracts (Tiwari *et al.,* 2009). The results are shown in Table 20.

Tableau 20. Minimum inhibitory concentrations (mg/ml) of dry bean polyphenolic extracts in liquid medium.

Strain	MICs (mg/ml)			
	PHRS	**PHRC**	**PHBS**	**PHBC**
Alternaria sp.	30	30	15	30
Aspergillus sp.	30	30	30	30
Fusarium sp.	30	30	30	30
Moniliella sp.	1,87	7,50	15	15
Penicillium sp.	15	30	15	30
Rhizopus sp.	15	30	15	30

The minimum inhibitory concentration (MIC), defined as the lowest concentration of polyphĕnols that resulted in the competitive inhibition of mycĕte growth (Jong *et al*., 2010), ranged from 1.87 to 30 mg/ml for healthy red bean polyphĕnolic extracts, from 7.5 to 30 mg/ml for contaminated red bean polyphĕnolic extracts and from 15 to 30 mg/ml for healthy white bean polyphĕnolic extracts and contaminated white bean polyphenolic extracts. Strains of the genus *Moniliella were found to be the* most sensitive fungal strains to the polyphenolic extracts studied, particularly those of healthy kidney beans (MIC=1.87 mg/ml).

8. 4. Fungistatic and fungicidal concentrations

The fungistatic (CFS) and fungicidal (CF) concentrations expressed in mg/ml are shown in Table 21.

Tableau 21. Fungistatic (CFS) and fungicidal (CF) concentrations in mg/ml of dry bean polyphenolic extracts on isolated strains.

Strain	CF PH	CFS [RS	CF PH	CFS [RC	CF PH	CFS [BS	CF PH	CFS BC
Alternaria sp.	30	ND	ND	30	30	15	30	ND
Aspergillus sp.	30	ND	30	ND	30	ND	30	ND
Fusarium sp.	30	ND	30	ND	30	ND	30	ND
Moniliella sp.	3,75	1,87	7,5	ND	15	ND	15	ND
Penicillium sp.	ND	15	ND	15	30	15	ND	30
Rhizopus sp.	15	ND	30	ND	30	15	ND	30

ND: Non Determines

The subcultures carried out after the MICs were obtained revealed varying levels of activity of the dried bean polyphenolic extracts on the strains tested. Healthy kidney bean polyphenolic extracts showed fungistatic activity on fungal strains at 15 mg/ml for *Penicillium sp.* and 1.87 mg/ml for *Moniliella sp.* Polyphenols from contaminated kidney beans and healthy white beans were fungistatic at 15 mg/ml on *Penicillium sp.* and at 30 mg/ml and 15 mg/ml respectively on *Alternaria sp.* Polyphenolic extracts of contaminated white beans were fungistatic on *Penicillium sp.* and *Rhizopus sp.* at 30 mg/ml.

According to the results of the antifungal activity of polyphënolic extracts of ground whole beans, reëaliseëed by the direct contact method, the strains ëtudiëes do not have the same sensitivity. This is thought to be due to the nature of the wall of fungal strains, which is composed of a complex network of proteins and polysaccharides and varies in composition depending on the fungal species (Yen and Chang, 2008). Disruption of this matrix by adsorption of phenolic compounds following polyphenol/polysaccharide membrane compiexation, can result in a reduction in the fluidity of the inner and outer layers of the wall, which becomes defective and susceptible to osmotic lysis by antifungal agents (Domineco *et al.,* 2005)

In addition, the antifungal activity of polyphenolic extracts could also be due to the fact that certain polyphenolic compounds bind to the proteins of microorganisms, blocking their enzymatic activities (Mohamed Sham *et al.*, 2010). In addition, DNA and RNA synthesis may be inhibited (Hadi, 2004).

According to Wiley (2005), the characteristic colour observed in many moulds is due to the pigmentation of conidiums. Consequently, the change in colour of *Aspergillus sp.*, *Fusarium sp.* and *Penicillium sp.* strains could be due to the effect of polyphenolic extracts from dry bean kernels on conidia. Several authors, including Sharma *and* Tripath (2007), have found that plant extracts can cause morphological changes including insufficient sporulation, loss of pigmentation, abnormal conidiophore development and hyphal deformation.

The toxicity of tannins, the most widely distributed phenolic group in beans, to moulds is a well-established fact, either through *in vitro* experiments reducing the growth of filamentous fungi, sometimes including blocking or slowing spore germination, or through biochemical approaches affecting mould respiration (Macheix *et al.*, 2005).

In addition to tannins, a large number of simple dry bean phenolic compounds (ferulic acid, cinnamics, flavonoids, etc.) also have antifungal activity. Here again, phenolic compounds can act either directly by blocking mould growth (Macheix *et al.*, 2005) or indirectly by inhibiting certain extracellular enzymes, which they use to penetrate grains, such as cellulases and pectinases (Chërif *et al., 2007)*. The same observations have ële made by several researchers (Table 22).

Table 22. Some phënolic composës active *in vitro* on certain mycëtes.

Phenolic compounds	Target moulds	References
Cinnamic acid	*Alternaria, Sclerotinia.*	(Lattanzio *et al.*, 2001)
fërulic acid	*Penicillium.*	(Lattanzio *et al.*, 2001)
Flavonoides	*Alternaria, Penicillium.*	(Chërif *et al.*, 2007)

The determination of minimum inhibitory concentrations, fungicidal and fungistatic concentrations, by the dilution method in liquid media, confirms the results obtained in solid media by the direct contact method. This antifungal activity^ seems to be Hëc to the presence of certain chemical functions, such as the presence of an aromatic ring and the hydroxyl group in the structure of phënolic compounds (Shirazdl *et al.*, 2011).

Conclusion and outlook

Dried vegetables are undoubtedly among the raw materials most exposed to fungal contamination. Fungal growth on these substrates can have several consequences: alteration of organoleptic properties, reduction in nutritional quality, outbreaks of disease or accumulation of toxic compounds. Polyphenols are among the constituents of grain that can play a role in its defence against biotic and abiotic stress. In recent years, they have attracted growing interest from the food industry.

This study was ë1.ë conducted with the aim of assessing the total polyphënol content of whole grains of two dry bean varieties (white and red), to investigate the main mould genera contaminating these beans, taking into account the intra-varietal effect (healthy and damaged samples of the same variety) and the inter-varietal effect (samples belonging to different varieties) and to test *in vitro* the antifungal effect of total polyphenols on isolated mould strains.

The first essential analysis was an assessment of the moisture content of the grains, which revealed higher values in batches thought to be contaminated. In general, these levels are favourable to the development of moulds, which are xerotolerant micro-organisms.

The mycological analysis carried out on our samples enabled us to isolate 43 strains from the two dry bean varieties analysed. The highest contamination rate was observed in the white variety, which appears to be more susceptible to fungal attack. Purification of the isolated strains enabled us to identify six mould genera, namely : *Alternaria, Aspergillus, Fusarium, Moniliella, Penicillium* and *Rhizopus,* with *Penicillium* dominating in both varieties. Strains belonging to this genus are referred to as storage moulds.

The results of the extraction and determination of total polyphenols showed that their content is not stable and varies greatly with the variety, being higher in the red variety than in the white variety. As for the intra-varietal difference, healthy samples are richer in total polyphenols than contaminated samples.

The search for possible correlations between the contamination rate, the moisture content and the total polyphenol content of the different dry bean samples indicates a positively significant correlation between the moisture content and the contamination rate of the beans. On the other hand, the contamination rate was significantly negatively correlated with the polyphenol content.

The antifungal test, carried out using the direct contact method, of the polyphenolic extracts against the six strains isolated and tested showed an antifungal effect. This effect varied according to the strain and the dose applied. Strains belonging to the *Moniliella* and *Alternaria* genera appeared to be the most sensitive, with all the polyphenolic extracts inhibiting their growth by more than 50%.

The dilution method confirmed the results of the direct contact method. Strains of the *Moniliella* genus proved to be the most sensitive fungal strains to the polyphenolic extracts studied, particularly those of the healthy red variety. These extracts showed fungistatic activity, but this varied according to the strain and concentration considered (15 mg/ml for *Penicillium sp.* and 1.87 mg/ml for *Moniliella sp.).* The polyphenolic extracts of the contaminated red variety and the healthy white variety were also fungistatic and varied according to the strain and concentration considered (15 mg/ml for *Penicillium sp.* and 30 mg/ml and 15 mg/ml respectively for *Alternaria sp.).* Polyphenolic extracts of the contaminated white variety are fungistatic on *Penicillium sp.* and *Rhizopus sp.* at 30 mg/ml.

All the concentrations of polyphenolic extracts applied partially or completely inhibited the growth of the moulds tested. The results obtained indicate that the most sensitive strains are *Alternaria sp.*, *Moniliella sp.* and *Rhizopus sp.*, while *Fusarium sp. Penicillium sp.* and *Aspergillus sp.* are the most resistant.

Following on from the present study, it would be interesting to carry out a more in-depth study to identify the species of strains isolated, quantify and characterise the nature of the polyphenolic compounds present in the various histological layers of the dry bean kernel, evaluate the antifungal power of each phenolic compound and examine the possibility of genetically improving the content of these polyphenolic compounds in order to improve resistance to fungal infestation.

References

Abad M-J., Ansuategui M. and Bermejo P., 2007. Active antifungal substances from natural sources, *ARKIVOC* (VII): 116-145.

Abdel Massih M., 2007. Moulds: identification, sources of contamination and control methods. *Pole technologique agroalimentaire*: 3.

Alavi S-H-R., Yassa N. and Fazeli M-R., 2005. Chemical constituents and antibacterial activity of essential oil of *Peucedanum ruthenicum* M. Bieb. Fruits. *SHR IJPS* 1 (4): 217222.

Alcamo E-I., 1984. Fundamentals of Microbiology. *Addison Wesly publishing company*, London: 310-341; 617-699.

Aparicio F-X., Manzo-Bonilla L. and Loarca-Pina G., 2005. Comparison of antimutagenic activity of phenolic compounds in newly harvested and stored common beans *Phaseolus Vulgaris* against aflatoxin B1. *Journal of the Science of Food* (70): 73-78.

Arkoyld W-R. and Daughty J., 1982. Legume seeds in the human diet. *FAO*: 152.

Ausubel F-M., 2005. Are innate immune signaling pathways in plants and animals conserved, *Nature Immunology* (6): 973-79.

Badillet G., De Brieve C. and Nielio E., 1987. Champignons contaminants des cultures, champignons opportunistes, Atlas clinique et biologique II, *VARIA*: 153-155.

Bahorun T., Gressier B., Trotin F., Brunet C., Dine T., Luyckx M., Vasseur J., Cazin M., Cazin J-C. et Pinkas M., 1996. Oxygen species scavenging activity of phenolic extracts from hawthorn fresh plant organs and pharmaceutical preparations. *Arzneiminittelforschung/Drug Research*. 46 II (11): 1086 - 1089.

Bajpai V-K., Shukla S. and Kang C-S., 2008. Chemical composition and antifungal activity of essential oil and various extract of Silene armeria L. *Bioresource Technology, Elsevier* (99): 8903-8908.

Bajpai V-K., Shukla S. and Kang C-S., 2010. Antifungal activity of leaf essential oil and extracts of *Metasequoia glyptostroboides*. *Journal. American. Oil. Chemistry. Soc* (87): 327-336.

Basset T., 2009. Biological degradation. *BNF. Bussy Saint Georges* : 04.

Batawita K., Kokon K., Akpagona K., Koumaglo K. and Bouchet P., 2002. Activite antifongique d'une espece en voie de disparition de la flore togolaise: *Conyza aegyptiaee*. Cite par Akroum S (2006), *memoire de magister*, universite de Constantine: 58.

Baudoin J-P., Crabbe J., Beatrice L. and Beatrice L., 2004. Interspecific hybridization with *Phaseolus vulgaris* L.: embryo development and its genetics. *Research Signpost*: 349364.

Baudoin J-P., Siliic S. and Jacquemin J-M., 2011. Use of induced mutations to study embryogenesis in *Phaseolus vulgaris* L. bean and two model plants, *Arabidopsis thaliana* and *Zea mays* L. *Biotechnology. Agronomy. Soc. Environment*: 195205.

Beninger C-W. and Hosfield G-L., 2003. Antioxidant activity of extracts, condensed tannin fractions and pure flavonoids from *Phaseolus vulgaris* L. seed coat color genotypes. *Journal of Agricultural and Food Chemistry* (51): 7879-7883.

Beta T., Nam S., Dexter J-E. and Sapirstein H-D., 2005. Phenolic content and antioxidant activity of pearled wheat and roller-milled fractions. *Cereal Chemistry*, 82(4): 90-393.

Boiron P., 1996. Organisation and biology of fungi. *Nathan*: 13-69.

Botton B., Breton A., Fevre M., Gauthier S., Guy P., Larpent J-P., Reymond P., Sanglier J-J., Vayssier Y. et Veau P., 1990. Useful and harmful moulds, industrial importance, *Masson*, Paris:

349P.
Bouayed J., Rammal H., Younos C., Dicko A. and Soulimani R., 2008. Caracterisation et bio ëvaluation des polyphënols: nouveaux domaines d'application en sante et nutrition. *Springer*. France : 4.
Bouchet P-H., Guignard J-L et Vihard J., 1999. Les champignons mycologie fondamentale et applique. *Masson*: 36-45.
Bouchet P., 2005. Les champignons: mycologie fondamentale et appliquee. *Masson*: 1- 23-25.
Boudra H., 2002. La contamination par les moisissures et les mycotoxines des fourrages conserves signification et prevention, *INRA*: 11.
Bourgeaois C-M. and Larpent L-P., 1996. Microbiologie Alimentaire, les fermentations alimentaires. Tome 2. *Lavoisier*: 17-19.
Bousseboua H., 2005. Elements of microbiology. *Campus club*, n°2: 14.
Braga F-G., Bouzada M-L-M., Fabri R-L., Matos M-O., Moreira F-O., Scio E. and Coimbra E-S., 2007. Antileishmanial and antifungal activity of plants used in traditional medicine in Brazil. *Journal of Ethnopharmacology* 111: 396-402.
Branger A., Richer M-C. and Roustel S., 2007. Alimentation, securite et microbiologie. *Educagri*, Dijon: 32-33.
Brochard G. and Le Bacle C., 2009. Mycotoxin in the workplace: origin and toxic properties of the main mycotoxins, medico-technical dossier n°119. *Departement etudes et assistance medicales, INRS*: 292.
Brownlee H-E., Hedger J. and Scott I-M., 1992. Effects of a range of procyanidins on the cocoa pathogen *Crinipallis perniciosa. Plant Pathology* (40): 227-232.
Bulter M-J. and Day A-W., 1998. Fungal melanins. *Canadian Journal Microbiology.* 44: 11151136.
Cahagnier B., Dragacc S., Frayssinet C., Fremy J-M., Hennebert G-L., Lesage-Meessen L., Multon J-L., Richard-Molard D. and Roquebert M-F., 1998. Moulds in foods with low hydës. Cite; by Nguyen M-T-M (2007). *These doctorat*, institut national polytechnique de Toulouse: 147P.
Cahagnier B. and Rhichard M-D., 1998. Analyse mycologique in Moisissures des aliments peu hydrates. *Technologie et Documentation*: 140-158.
Cahagnier R-B., 1996. Cereales et produits derives, *Tec et Toc, Lavoisier*, Paris: 392-413.
Cairns-Fuller V., Aldred D. and Magan N., 2005. Water, temperature and gas composition interactions affect growth and ochratoxin A production by isolates of *Penicillium verrucosum* on wheat grain, *Journal of Microbiology*: 1215-1221.
Cardador-Martinez A., Loarca-Pina G. and Oomah B-D., 2002. Antioxidant activity in common beans (*Phaseolus vulgaris* L.). *Journal of Agricultural and Food Chemistry* (50): 69756980.
Castegnaro M. and Pfohl-Leszkowicz A., 2002. Balkan endemic nephropathy and the associated urinary tract tumours: review on etiological causes, potential role of mycotoxins. *Food Additive and Contaminants* . 19 (3): 282-302.
Catanzariti A-M., Dodds P-N. and Ellis J-G., 2007. Avirulence proteins from haustoria-forming pathogens. *FEMS Microbiology. Lett*; 269: 88-181.
Cazaux M., 2009. Study of the resistance of the model legume *Medicago truncatula* to *Colletotrichum trifolii*, agent of anthracnose. *These de doctorat*. University of Toulouse: 178 P.
Chabasse D., Bouchra J-P., De Gentile L., Brun S., Cimon B. and Penn P., 2002. Les moisissures d'interet medical, cahier de formation N° 25, *Bioforma,* Paris: 160.

Chahardehi A-M., Ibrahim D. and Sulaiman S-F., 2010. Antioxidant, antimicrobial activity and toxicity test of *Pileamicrophylla. International Journal of Microbiology*: 6.
Champion R., 1997. Identifying seed-borne fungi. *INRA*: 20-26.
Chang C-W., Chang W-L., Chang S-T. and Cheng S-S., 2008. Antibacterial activities of plant essential oils against *Legionella pneumophila. Water Research* 42: 78-286.
Charles M. and Benbrook P-D., 2005. Increasing the antioxidant content of foods through organic farming and food processing. State of the Science Report. *The Organic Center*: 10.
Charpentier J-P. and Boizot N., 2006. Mëthode rapide devaluation du contenu en composës phënoliques des organes d'un arbre forestier. AтёHoгaйoп gënëtique et physiologie forestieres. *INRA*: 79.
Chërif M., Arfaoui A. and Rhaiem A., 2007. Phenolic compounds and their role in biocontrol and resistance of chickpea to fungal pathogenic attacks. *Tunisian Journal of Plant Protection* (2): 7-21.
Chisholm S-T., Coaker G., Day B. and Staskawicz BJ., 2006. Host-microbe interactions: shaping the evolution of the plant immune response. *Cell* (124): 803-14.
Christensen C-M., 1994. Mycology of soft ЫЁ: СИЫПё grain technology and consëquences on finished products. Vol 7. N° 1: 27-32.
Christensen C-M., Meronuk R-A. and Sauer D-B., 1982. Micoflora (9), *American association of cereal chemists*: 219-240.
CIAT (International Center for Tropical Agriculture), 2003. New bean variëtës for ëthiopian farmers. *Highlights*: 2.
Cowan M-M., 1999. Plant Products as Antimicrobial Agents. *Clinical Microbiology Reviews*. 12 (4): 564-582.
Cronk Q., Ojeda I. and Pennington R-T., 2006. Legume comparative genomics: progress in phylogenetics and phylogenomics. *Cur Opin Plant Biol* 9: 99-103.
Cruz J-F. and Diop A., 1989. Gënie agricole et dëveloppement, techniques d'entreposage. *FAO*, Italy: 7.
Cup J-L. and Legnaud-Rouand C., 1992. Les graines de tegumineuses, *ESF*, Paris: 941-96.
D'Halewyn M-A., Leclerc J-M., King N., Bëlanger M., Legris M. and Frenette Y., 2002. Les risques a la santë asso^s a la presence de moisissures en milieu intërieur, Document synthese Tire de : Les risques a la santë asso^s a la presence de moisissures en milieu intërieur, *Institut national de santepublique du Quebec*: 16.
Daniel G., Debouk I-N-G. and Rigoberto H., 1987. Morphologie de la plante du haricot sec (*Phaseolus vulgaris* L.), *CIAT*: 56.
De Aguirre L., Hurst S-F., Choi J-S., Shin J-H., Hinrikson H-P. and Morrison C-J., 2004. Rapid differentiation of *Aspergillus* species from other medically important opportunistic molds and yeasts by PCR-enzyme immunoassay, *J. Clin. Microbiol.* 42 (8): 3495-3504.
De Lucia M. and Assennato, 1992. Grain post-harvest. *FAO*. Rome: 72.
Debete J-M., 2005. Phytochemical and pharmacological study of *Cassia nigricans* Vahl (*Caesalpiniaceae*) usedë in the treatment of dermatoses in Chad. *PhD thesis*. Universe de Bamako: 316 P.
Dehimat L., 1990. Etude des mycotoxines sëcrëtëes par les moisissures contaminant les cërëales stockëes dans la region de 1 est ttigerien. *These magister INS*. Uni versite de Constantine: 128.
Derwich E., Benziane Z. and Boukir A., 2010. GC/MS Analysis and antibacterial activity of the essential oil of *Mentha pulegium* grown in Morocco. *Res. Journal of Agricultural and Biology*. 6 (3):191-198.

Dinelli G., Bonetti A., Minelli M., Marotti I., Catizone P. and Mazzanti A., 2006. Content of flavonols in italian bean *(Phaseolus vulgaris L.)* ecotypes. *Food Chemistry (*99): 105-114.
Dixon R-A. and Paiva N-L., 1995. Stress induced phenylpropanoid metabolism, *Plant cell* (7): 1085-1097.
Djebali N., 2008. Study of resistance mechanisms in the modeled plant *Medicago truncatula* against two major pathogens of cultivated legumes: *Phoma medicaginis* and *Aphanomyces euteiches. These de doctorat*. University of Toulouse. 209 P.
Domenico T., Francesco C., Maria G-S., Vincenza V., Mariateresa C-D., Antonella S., Gabriela M. and Giuseppe B., 2005. Mechanisms of antibacterial action of three monoterpenes. *Antimicrobial Agents and Chemotherapy,* 49: 2474-2478.
Doumandji A., Doumandji S. and Doumandji M-B., 2003. Le stockage et la lutte contre les ennemis des cereales. *Seminar on flour milling and cereal industries*: 4-14.
Drevon J-J., 2009. Cooperation on the theme: Phosphorus use efficiency and symbiorhizobial nitrogen fixation in bean cultivation (within the framework of the Tassili programme N° 08MDU721), *Compte rendu de missions, INRA*: 6.
Drewnowski A. and Gomez-Carneros C., 2000. Bitter taste, phytonutrients, and the consumer: a review. *American Journal of Clinical Nutrition* (72): 1424-1435.
DSASI, 2009. Direction des statistiques agricoles et des systemes d'information, Superficie et production des legumes secs en Algerie (2006-2009).
Duron L., 1999. Le transport maritime des produits cerealiers, *memoire pour dess*, université Marseille: 81.
Ebrahimi N-S., Hadian J., Mirjalili M-H., Sonboli A. and Yousefzadi M., 2008. Essential oil composition and antimibacterial activity of *Thymus caramanicus* at different phonological stages. *Food chemistry* (110) : 927-931.
Ekoumou C., 2003. Phytochemical and pharmacological study of five traditional recipes used in the treatment of urinary tract infections and cystitis. *These doctorat*. University of Bamako: 168.
Esekhiagbe M., Uzuazokaro Agatemor M-M. and Agatemor C., 2009. Phenolic content and antimicrobial potentials of *Xylopia aethiopica* and *Myristica argentea. Macedonian Journal of Chemistry and Chemical Engineering*, Vol. 28, No. 2: 159-162.
FAO and WHO, 2007. Codex alimentarius: Cëréales, legumes, pulses and vegetable protein. *Joint FAO/WHO programme on codex standards for selected pulses*. Rome: 13-14.
FAO, 1984. Loss of quality of food grains after harvesting, *FAO food and nutrition study*. Rome: 17.
FAO, 1994. FAO study, Forets. *Forest service*, Rome: 12-16.
Feillet P., 2000. Le grain de ble composition et utilisation. *INRA*, Paris: 53.
Feliachi K., 2006. Rapport national sur l'etat des Ressources phytogenetiques pour L'alimentation et l'agriculture*, INRAA*: 68.
Ferrari J., 2002. Contribution a la connaissance du metabolisme secondaire des *Thymelaeaceae* et investigation phytochimique de l'une de elles: *Gnidia involucrata Steud.* ex A. Rich. *These doctorat*. University of Lausanne: 320 P.
Flamini G. and Luigi Cioni P., 2003. Activity of plant extracts, essential oils, and pure compounds against fungi contaminating foodstuffs and causing infections in human beings and animals: A six-year experience (1995-2000). *Food products press: Crop Science*. New York: 279-297.
Fleuriet A. and Macheix J-J., 2003. Phenolics acids in fruits and vegetables. *Marcel Dekker*,

New York: 1-41.
Francois V., 2004. Determination d'indicateurs d'acceleration et de stabilisation de deterioration des cereales, *PhD thesis*, University of Limoges. 360 P.
Galtier P., Loiseau N., Oswald I-P. and Puel O., 2005. Toxicologie des mycotoxines: dangers et risques en alimentation humaine et animale Bull. *Academy. Veterinary*, Tome 159, N°1. France: 12-15.
Garcia-Salas P., Morales-Soto A., Segura-Carretero A. and Fernandez-Gutierrez A., 2010. Phenolic-compound-extraction systems for fruit and Vegetable samples. *Molecules* (15): 8813-8826.
Gepts P., Beavis WD, Brummer EC, Shoemaker RC, Stalker HT, Weeden NF and Young ND, 2005. Legumes as a model plant family. Genomics for food and feed report of the cross-legume advances through genomics conference. *Plant Physiology* 137: 1228-1235.
Godon B. and Loisel W., 1997. Guide pratique d'analyses dans les industries des cereales, *Technologie et Documentation*, Paris: 819 P.
Goix J., 1986. The bean bruchid. *Revue Phytoma- Defense des cultures*: 48-49.
Golam A., Khandaker L., Berthold J., Gates L., Peters K., Delong H and Hossain, 2011. Anthocyanin, total polyphenols and antioxidant activity of common bean. *American. Journal of Food and Technology*. 6: 385-394.
Gonzalez de Mejia E., Castano-Tostado E. and Loarca-Pina G., 1999. Antimutagenic effects of natural phenolic compounds in beans. *Mutation Research* (441): 1-9.
Graham P-H. and Vance CP., 2003. Legumes: Importance and constraints to greater use. *Plant Physiology*, 131: 872-877.
Granito M., Paolini M. and Përez S., 2007. Polyphenols and antioxidant capacity of *Phaseolus vulgaris* stored under extreme conditions and processed. *LWT Food Science and Technology* (41): 994-999.
Guiraud J-P., 2003. Food microbiology. *Duond*, Paris. 651 P.
Hadi M., 2004. La quercëtine et ses dërivës: molecules a caractere pro- oxydant ou capteurs de radicaux libres; études et applications therapeutiques. *These doctorat*. Universite Louis Pasteur: 155 P.
Hageskal G., Knutsen A-K., Gaustad P., de Hoog G-S. and Skaarl I., 2006. Diversity and significance of mold species in Norwegian drinking water, *Appl. Environment. Microbiology*. 72 (12): 7586-7593.
Hahlbrock K., Scheel D., Logemann E., Nurnberger T., Parniske M., Reinold S., Sacks W-R. and Schmelzer E., 1995. Oligopeptide elicitormediated defense gene activation in cultured parsley cells. *Proc Nalt Acad Sci USA*, 92: 4150-4157.
Hale A-L., 2003. Screening Potato Genotypes for Antioxidant Activity, Identification of the Responsible Compounds, and Differentiating Russet Norkotah Strains Using Aflp and Microsatellite Marker Analysis. *Office of Graduate Studies of Texas A&M University. Genetics*: 260 P.
Halliwell B. and Gutteridge J-M-C., 1995. The definition and measurement of antioxidants in biological systems. *Free Radical Biology and Medicine*, 18: 125-126.
Harborne J-B., 1989. General procedures and measurement of total phenolics. *Plant phenolics. Academic Press,* London: 1-28.
Hawksworth D-L., Kirk P-M., Sutton B-C. and Pegler D-N., 1995. Ainsworth and bisby's dictionary of the fungi.*CAB International*, Unitd Kingdom: 616.
Hayma J., 2004. Storage of tropical agricultural products, *Agromisa Foundation*, Wageningen;

80.
Hinrikson H-P., Hurst S-F., De Aguirre L. and Morrison C-J., 2005. Molecular methods for the identification of *Aspergillus* species, 43 (1): 129-137.
Horbowicz M., Kosson R., Grzesiuk A. and D^bski H., 2008. Anthocyanins of fruits and vegetables, their occurrence, analysis and role in human nutrition. *Vegetable Crops Res. Bull*: 5-22.
Hossain M-A., Zhari I., Atiqur R. and Chul Kang S., 2008. Chemical composition and anti-fungal properties of the essential oils and crude extracts of *Orthosiphon stamineus* Benth. *Industrial crops and products* (27): 328-334.
Huang D., Ou B. and Prior R-L., 2005. The Chemistry behind Antioxidant Capacity Assays. *Journal of Agricultural & Food Chemistry*. 53: 1841-1856.
Hubert J., Stejskal V., Munzbergova Z., Kubatova A., Vanova M. and Zd'arkova E., 2007. Mites and fungi in heavily infested stores in the Czech Republic, *J. Econ. Entomol.* 97(6): 21442153.
Hussin N-M., Muse R., Ahmad S., Ramli J., Mahmood M., Sulaiman M-R, Shukor M-A-Y., Rahman M-F-A. et Aziz K-N-K., 2009. Antifungal activity of extracts and phenolic compounds from *Barringtonia racemosa* L. (*Lecythidaceae*). *African Journal of Biotechnology* Vol. 8 (12): 2835-2842.
Jard N., 1995. Les maladies des grains- Tome I. *Universite. OmarMokhtar*, LYBIA: 517-522.
Jones G-A., Mcallister T-A., Muir A-I-D. and Cheng K-J., 1994. Effects of Sainfoin (*Onobrychis viciifolia* Scop.) Condensed tannins on growth and proteolysis by four strains of ruminal bacteriat. *Applied and environmental microbiology* (60) 4: 1374-1378.
Jones J-D-G. and Dangl J-L., 2006. The plant immune system. *Nature*; 444: 323-29.
Jong H., Bruce C., Campbell N., Chan K-L. and Russell J-M-A., 2010. Augmenting the activity of antifungal agents against aspergilli using structural analogues of benzoic acid as chemosensitizing agents. *Bfungal biology* 114: 817-824.
Kassemi N., 2006. Relationship between a phytophagous insect and its main host plant: the case of the bean bruchid (*Acanthoscelides obtectus*)(*Coleoptera bruchidae*). *These magister*. Universite de Tlemcen: 107 P.
Kawamura F., Shaharuddin N-A., Sulaiman O., Hashim R. and Ohara S., 2010. Evaluation on antioxidant activity, antifungal activity and total phenols of 11 selected commercial Malaysian timber species. *JARQ* 44 (3): 319-324.
Keller S-E., Sullivan T-M. and Chirtel S., 1997. Factors affecting the growth of *Fusarium proliferatum* and the production of fumonisin B1: oxygen and pH, *Indust. Microbiol, Biotechnology*, 19: 305-309.
Kellouche A. and Soltani N., 2005. Activite biologique des poudres de cinq plantes et de 1'huile essentielle d'une d'elles sur *Callosobruchus maculatus, International Journal of Tropical Insect Science* Vol. 24, No. 1: 184-191.
Khallil A-R-M., 2001. Phytofungitoxic Properties in the aqueous extracts of some plants. *Pakistan Journal of Biological Sciences*. 4 (4): 392-394.
Krogh P., 1987. Ochratoxin A in food. Mycotoxin in food. *Academic Press*, San Diego: 97121.
Kuhnie G., 2003. Investigation of flavonoids and their in vivo metabolite forms using tandem mass spectrometry. *Marcel Dekker*, New York: 145-163.
Laib I., 2009. Study of the antioxidant and antifungal activities of the essential oil of the dried flowers of *Lavandula officinalis* on the moulds of dried vegetables. *These magistere*. INATAA, Constantine: 124 P.

Laparra, J-M., Glahn R-P. and Miller D-D., 2008. Bioaccessibility of phenols in common beans (*Phaseolus vulgaris* L.) and iron (Fe) availability to caco-2 cells. *J. Agric. Food Chem.* 56: 10999-11005.
Larpent J-P. and Larpent G-M., 1990. Memento technique de microbiologie. *Tec et Doc, Lavoisier*, Paris: 78-79-81.
Lattanzio V., Di Venere D., Linsalata V., Bertolini P., Ippolito A. and Salerno M., 2001. Low temperature metabolism of apple phenolics and quiescence of *Phlyctaena vagabunda. Journal of Agriculture and Food Chemistry 49*: 5817-5821.
Lee K-W., Kim Y-J., Lee H-J. and Lee C-Y., 2003. Cocoa has more phenolic phytochemicals and a higher antioxidant capacity than teas and red wine. *Journal of Agricultural and Food Chemistry*. 51: 7292-7295.
Leighton T., Ginther C., Fluss L., Harter W-K., Cansado J. and Notario V., 1992. Molecular characterization of quercetin and quercetin glycosidesvin Allium vegetables: their effects on malignant cell transformation. *American Chemical Society*, Washington: 220-238.
Lin L-Z., Harnly J-M., Pastor-Corrales M. and Luthria D-L., 2008. The polyphenolic profiles of common bean (*Phaseolus Vulgaris* L.). *Food Chemistry*, 107: 399-410.
Lis-Balchin M., 2002. Lavender: the genus *Lavandula. Taylor and Francis*, London: 37-200.
Lugasi A., Hovari J., Sagi K-V. and Bиo L., 2003. The role of antioxidant phytonutrients in the preventıon of dıseases. *Acta Biologica Szegediensis*. 47(1-4); 119-125.
Luthria D. and Pastor-Corrales M., 2006. Phenolic acids content of fifteen dry edible bean (*Phaseolus Vulgaris* L.) varieties. *Journal of Food Comp. Anal*: 205-211.
M.A (ministere de ramenagement du temtoire et de l'environnement), 1998. L'agriculture par les chiffres, *Direction des statistiques agricoles et des enquetes economiques.*
Macarde C., 2004. Structural and ëvolutive analysis of a cluster of resistance gene analogues in *Phaseolus vulgaris. Memoire du diplôme de Γ ecole pratique des hautes etudes*. Universite Paris Sud: 31 P.
Macheix J-J., Fleuriet A. and Billot J., 1990. Fruit phenolics, *CR Press*, Boca Raton: 378 P.
Macheix J-J., Fleuriet A. and Sarni Manchado. 2005. Phenolic compounds in plants: structure, biosynthesis, distribution and roles. In: les polyphenols en agroalimentaire, Cheynier V., Sarni Manchado P. *Lavoisier*, Paris: 510 P.
Madi A., 2010. Characterisation and comparison of the polyphenolic content of two medicinal plants (Thyme and Sage) and evidence of their biological activities. *These magister*. Universite Mentouri Constantine: 116 P.
Magan N. and Olsen M., 2004. Mycotoxins in food: Detection and control, *Woodhead Publishing in Food Science and Technology*: 190-203.
Manach C., Scalbert A., Morand C., Remesy C. and Jimenez L., 2004. Polyphenols: food sources and bioavailability. *American. Journal Clin Nutr*. 79(5): 727-747.
Markham K. and Bloor S., 1998. Analysis and identification of flavonoids in practice. In: flavonoids in health and disease, Rice-Evans C., Packer L. *Marcel Dekker*, New York: 134.
Meyer A., Deiana J. and Bernard A., 2004. Cour de microbiologie generale. *J.Appl.Microbiol.* 66 (4): 1523-1526.
Miliauskas G., Venskutonis P-R. and Van Beek T-A., 2004. Screening of radical scavenging activity of some medicinal and aromatic plant extract. *Food chemistry*, 85: 231-237.
Mishra A-K. and Dubey N-K., 1994. Evaluation of Some Essential Oils for Their Toxicity against Fungi Causing Deterioration of Stored Food Commodities. *Applied and Microbiology*. 60 (4): 1101-1105.

Mohamed Sham S., Hansi H., Priscilla D. and Kavitha T., 2010. Antimicrobial activity and phytochemical analysis of selected Indian folk medicinal plants. *International Journal of Pharma Sciences and Research (IJPSR)* Vol. 1(10): 430-434.

Mohammedi Z., 2006. Etude du pouvoir antimicrobien et antioxydant des huiles essentielles et flavonoi'des de quelques plantes de la region de Tlemcen. *These magistere*. Universite de Tlemcen: 104 P.

Moller B. and Herrmann K., 1982. Analysis of quinic acid esters of hydroxycinnamic acids in plant material by capillary gas chromatography and high performance liquid chromatography, *Journal of chromatography* 241: 371-379.

Moreau C., 1996. Moulds. In Bourgeois C-M., Mescle J-F and Succa J., 1996. Micobiologie alimentaire. Tome 1. *Lavoisier*. Paris: 234-235.

Mujica M-V., Granito M. and Soto N., 2009. Importance of the extraction method in the quantification of total phenolic compounds in *Phaseolus vulgaris* L.Venzuela. *Interciencia,* Vol. 34, N° 9: 650-654.

Multon J-L., 1982. Conservation et stockage des grains et des graines et produits dërivës (Cërëales, Oleagineux, Proteagineux, aliments pour animaux), *Lavoisier,* Paris: 576.

Navarro J-M., Flores P., Garrido C. and Martinez V., 2008. Changes in the contents of antioxidant compounds in pepper fruits at different ripening stages, as affected by salinity. *Food Chemistry*. 96: 66-73.

Nguyen M-T., 2007. Identification of mould species, potentially producing mycotoxins in rice marketed in five provinces of the central region of Vietnam- study of conditions that can reduce mycotoxin production. *These doctorat*. Institut national polytechnique de Toulouse: 147 P.

Nicklin J., Graeme-Cook K., Paget T. and Killington R., 2000. The essentials of microbiology. *Berti*: 210-216.

Nurnberger T. and Kemmerling B., 2006. Receptors protein kinases-Pattern recognition receptors in plant immunity. *Trends Plant Sci* (11): 519-22.

Oomah B-D., Cardador-Martinez A. and Loarca-Pina G., 2005. Phenolics and antioxidative activities in common beans (*Phaseolus vulgaris* L.). *Journal of the Science of Food and Agriculture*, 85: 935-942.

Orturno A., Baidez A., Gomey P. and Arenas M-C., 2005. *Citrus perasidi* and *Citrus sinensis* flavonoids: Their influence in the defense mechanism against *Penicillium digitatum*, cited by Akroum S (2006). *These magister*. Universite de Constantine: 81 P.

Ownagh A., Hasani A., Mardani K. and Ebrahimzadeh S., 2010. Antifungal effects of thyme, agastache and satureja essential oils on *Aspergillus fumigatus, Aspergillus flavus* and *Fusarium solani. Veterinary Research Forum*. 2: 99-105.

Parr A-J. and Bolwell P-G., 2000. Phenols in the plant and in man. The potential for possible nutritional enhancement of the diet by modifying the phenols content or profile. *Journal of the Science of Food and Agriculture*, 80: 985-1012.

Peterson S-W., 2006. Multilocus sequence analysis of Penicillium and Europenicillium species, *Rev. Iberoam Micol*, 23(3): 134-8.

Pfohl-Leszkowicz A., 1999. Les mycotoxines dans l'alimentation, ëvaluation et gestion des risques, *Lavoisier*, Paris: 478 P.

Pfohl-Leszkowicz A., 2001. Mycotoxins in food: ë Risk assessment and management. *Tec et Doc*: 3-14.

Phattayakorn K. and Wanchaitanawong P., 2009. Antimicrobial activity of thai herb extracts against coconut milk spoilage microorganisms. *Kasetsart J. Nat. Sci.* 43: 752-759.

Pietta P., Gardana C. and Pietta A., 2003. Flavoniods in herbs. In: Flavonoids in health and desease, Rice Evans C., Packer L. *Marcel Dekker*, New York: 43-69.
Pitt J-I. and Miscamble B-F., 1995. The normal mycoflora of commodities from Thailand. Beans, rice, small grains and other commodities. *International Journal of Food Microbiology* 23: 35-53.
Psotova J., Lasovsky J. and Vicar J., 2003. Metal-Chelating Properties, Electrochemical Behavior, Scavenging and cytoprotective activities of six natural phenolics. *Biomed.* 147(2): 147153.
Queiroz-Monici K-S., Costa G-E-A., Da-Silva N., Reis S-M-P-M. and De-Oliveira A-C., 2005. Bifidogenic effect of dietary fiber and resistant starch from leguminous on the intestinal microbiota of rats. *Nutrition,* 21: 602-609.
Ramiro D., 2009. Characterisation of resistance mechanisms involved in the responses of cafëier (*Coffea arabica*) to the orange rust agent (*Hemileia vastatrix*). *These doctorat,* Montpellier supagro: 230 P.
Reboux G., Bellanger A-P., Roussel S., Grenouillet F. and Millon L., 2010. Revue des Maladies Respiratoires, stirie " pollution de l'air inttirieur ". Mould and housing: health risks and the risks involved. *Elsevier Masson*, France. (27): 169-179.
Reynoso C., Ramos G. and Loarca P., 2006. Bioactive components in common beans (*Phaseolus vulgaris* L.). *Research Signpost* (2). India: 37-661.
Rhodes M-J-C., Wooltorton L-S-C. and Hill A-C., 1981. Changes in phenolic metabolism in fruit and vegetable tissues under stress. *Academic Press*, London: 193-220.
Riba A., Sabeau N., Mathieu F. and Lebrihi A., 2005. Premitires investigations sur les champignons producteurs d'Ochratoxine A dans la filitire ctirtiale en Algtirie. *Euro-Maghreb symposium on biological and chemical contaminants and food safety*, Ftis.
Ribtireau-Gayon P., 1968. Les compostis phtinoliques des vtigtitaux. *Dunod*, Paris: 254 P.
Rice-Evans C-A., Miller N-J., Bolwell P-G., Bramley P-M. and Pridham J-B., 1995. The relative antioxidant activities of plant-derived polyphenolic flavonoids. *Free Radical Research*, 22: 375-383.
Riviere J., 1975. Les applications industrielles de la microbiologie. Collection sciences agronomiques. *Masson et Cie*: 31-195.
Roquebert M-R., 1997. Moulds in nature, biology and contamination. Scientific publication of Vhisc'um National d'Histoire Naturelle, Laboratoire de Cryptogamie, Paris. Website: http://www.roquber@mnhn.fr.
Ross K-A., Beta T. and Arntfield S-D-A., 2009. Comparative study on the phenolic acids identified and quantified in dry beans using HPLC as affected by different extraction and hydrolysis methods. *Food Chemistry.* 113: 336-344.
Satish S., Raghavendra M-P., Mohana D-C. and Raveesha K-A., 2010. In vitro evaluation of the antifungal potentiality of *Polyalthia longifolia* against some sorghum grain moulds. *Journal of Agricultural Technology*. Vol.6(1): 135-150.
Scriban R., 1993. Biotechnology. *Lavoisier*: 32-690.
Serrano M., Zapata P-J., Castillo S., Guilten F., Martinez-Romero D. and Valero D., 2010. Antioxidant and nutritive constituents during sweet pepper development and ripening are enhanced by nitrophenolate treatments. *Food Chemistry.* 118: 497-503.
Sharma N. and Tripath A., 2007. Effects of *Citrus* (L.) Osbeck epicarp essential oil on growth and morphogenesis of *Aspergillus niger* (L.) Van Tieghem. *Microbiol. Res. in press*, 345-545.
Shirzad1 H., Hassani1 A., Ghosta Y., Abdollahi A., Finidokht R. and Meshkatalsadat M-H.,

2011. Assessment of the antifungal activity of natural compounds to reduce postharvest gray mould (*botrytis cinerea* pers.: fr.) Of kiwifruits (*actinidia deliciosa*) during storage. *Journal of plant protection research* Vol 51, N°1.

Singleton V-L., Orthofer R. and Lamuela-Raventos R-M., 1999. Analysis of total phenols and other oxidant substrates and antioxidants by means of Folin-Ciocalteu reagent. *Methods Enzymol*, 299: 152-178.

Siret C., 2000. Structure des aliments. *Techniques d'mgenieur*: 11.

Smith R., 2002. Fungal identification guide. Department of veterinary pathology, *Texas university*: 24-26.

Solis-Pereira S., Ernesto F-T., Gustero V-G. and Mariano-Gutierrez R., 1993. Effect of different carbon sources on the synthesis of pectinase by Aspergillus niger in submerged and solid state fermentation. *Appl. Microbiol. Biotechnol.* 39: 36-41.

Sprent JI, and Parsons R., 2000. Nitrogen fixation in legume and non legume trees. *Field crop Resh*, 65: 183-196.

Subrahmanyam M., Hemmady A. and Pawar S-G., 2001. Antibacterial Activity Of HoneyOn Bacteria Isolated From Wounds. *Annals of Burns and Fire Disasters*. XIV (I).

Sud D., Sharma O-P. and Sharma P-N., 2005. Seed mycoflora in kidney bean (*Phaseolus vulgaris L.*) in Himachal Pradesh. *Seed Res* 33: 103-107.

Suszka B., Muller C. and Bonnet Masimbert M., 1994. Forest hardwood seeds: from harvest to sowing. *Institut National de la Recherche Agronomique (INRA)*. Paris: 30-31.

Swinny E. and Markham K., 2003. Application of flavonoid analysis and identification techniques: isoflavones (phytoestrogens) and 3-deoxyanthocyanins. In: Flavonoids in health and disease, Rice Evans C., Packer L. *Marcel Dekker*, New York: 97-122.

Tabuc C., 2007. Fungal flora of different substrates and optimal conditions for mycotoxin production. *These doctorat*. Toulouse National Veterinary School. 190 P.

Tahani N., Elamrani A. Serghini-Caid H., Ouzouline M. and Khalid A., 2008. Isolation and identification of toxigenic mould strains. *Rev. Microbiol. Ind. San et Environn.* Vol2, N°1: 81-91.

Tameling W-I-L. and Takken F-L-W., 2008. Resistance proteins: scouts of the plant innate immune system. *Eur. J. Plant Pathol*: 243-55.

Tatsadjieu N-L., Jazet Dongmo P-M., Ngassoum M-B., Etoa F-X. et Mbofung C-M-F., 2009. Investigations on the essential oil of *Lippia rugosa* from Cameroon for its potential use as an antifungal agent against *Aspergillus flavus*. *Food Control.* 20:161-166.

Taylor J-W., 1996. Ascomycota Sac Fungi Department of Plant and Microbial Biology. *University of California Berkeley*: 94720-3120.

Thiam A-A., Drapron R. and Rhichard molard D., 1976. Causes of alteration of millet and sorghum flours. Ann. *Technol. Agric.*25 (3): 253-271.

Tiwari B-K., Valdramidis V-P., O'Donnell C-P., Muthukumarappan K., Bourke P. and Cullen P- J., 2009. Application of natural antimicrobials for food preservation. *Journal of Agricultural and Food Chemistry.* 57: 5987-6000.

Tseng T-C., Tu J-C. and Tzean S-S., 1995. Mycoflora and mycotoxins in dry bean (*Phaseolus vulgaris*) produced in Taiwan and in Ontario, Canada. *Bot. Bull. Acad. Sinica* 36: 229234.

Van Sumer C-F., 1989. Phenols and phenolics acids. In: Plant phenolics. Methods in plant biochemistry, vol 1, Harborne J-B. *Academic Press*, London: 29-73.

Vance C-P., Graham P-H. and Allan D-L., 2000. Biological nitrogen fixation. Phosphorus: a critical future need. *In*, Pedrosa F.O., Hungria M., Yates M.G., Newton W.E. Nitrogen fixation:

from Molecules to crop productivity. *Kluwer Academic Publishers,* New York : 839-867.
Wiley J., 2005. Essential microbiology. *Southern Gate, Chichester, West Sussex* PO19 8SQ, England: 481 P.
Wulf L-W. and Nagel C-W., 1978. High pressure liquid chromatographic separation of anthocyanins of *Vitis vinifera, Am. J. Enol. Vatic.* 29: 42-49.
Xu B. and Chang S., 2009. Total phenolic, phenolic acid, anthocyanin, flavan-3-ol and flavonol profiles and antioxidant properties of pinto and black beans (*Phaseolus vulgaris* L.) as affected by thermal processing. *Journal of Agricultural and Food Chemistry*, 57: 47544764.
Yen T-B. and Chang S-T., 2008. Synergistic effects of cinnamaldehyde in combination with eugenol against wood decay fungi. *Bioresource Technology* 99: 232-236.
Ying Tan S., Chi Kong Y., Elad T., Raymond P-G., Ross M-W., Xingen L. and Dennis D-M., 2008. Iron bioavailability to piglets from red and white common beans (*phaseolus vulgaris*). *J. Agric. Food Chem*: 5008-5014.
Yrjonen T., 2004. Extraction and Planar Chromatographic Separation Techniques in the Analysis of Natural Products. Conference *Room 513 at Viikki Infocentre (Viikinkaari 11), Faculty of Pharmacy of the University of Helsinki*: 64.
Zarrin M., Amirrajab N. and Nejad B-S., 2010. *In vitro* antifungal activity of satureja khuzestanica jamzad against *Cryptococcus neoformans*. *Pak. J. Med. Sci*; 26 (4): 880-882.
Zem T-L. and Fernondez M-L., 2005. Cardioprotective effects of dietary polyphenols. *The Journal of Nutrition*. 135: 2291-2294.

Appendices

Appendix 01. Composition of culture media (Guiraud, 2003).

PDA environment

Potato extract 1000ml

Glucose 20g

Agar 20g

This medium is autoclaved at 120°C for 15 min.

pH= 4.5

AMANN Lactophenol

Phënol pure.. cH^alШзё20g

Lactic acid 20g

.... Glycërine40g

.... Distilled water20ml

Store in coloured bottle away from light.

Appendix 02. Optical census of polyphënolic extracts.

Polyphenolic extracts	**PHRS**	**PHRC**	**PHBS**	**PHBC**
Optical densities	0,55±0,03	0,52±0,01	0,46±0,07	0,37±0,01

Appendix 03. Inhibition effect of polyphdnolic extracts against *Alternaria sp*. strain (a): PHRS, (b): PHRC, (c): PHBS, (d): PHBC

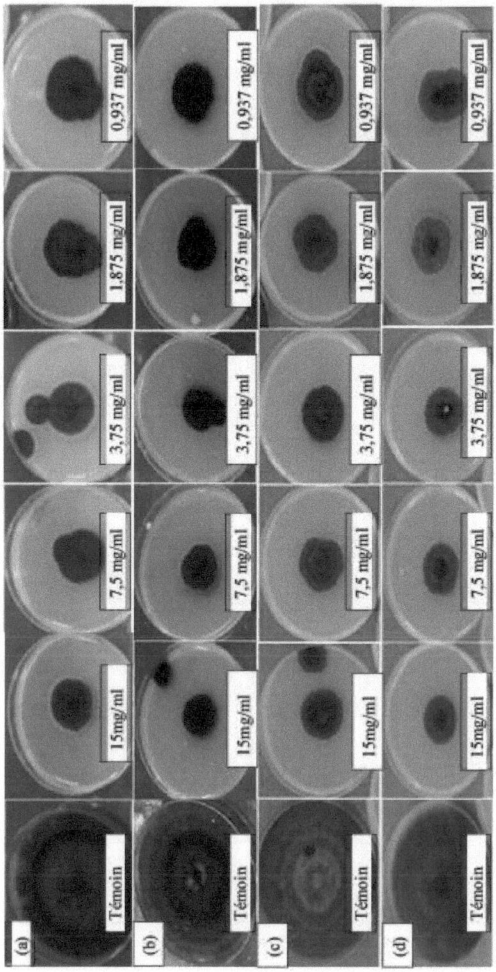

Appendix 04. Inhibition effect of polyphdnolic extracts against *Fusarium sp.* strain (a): PHRS, (b): PHRC, (c): PHBS, (d): PHBC.

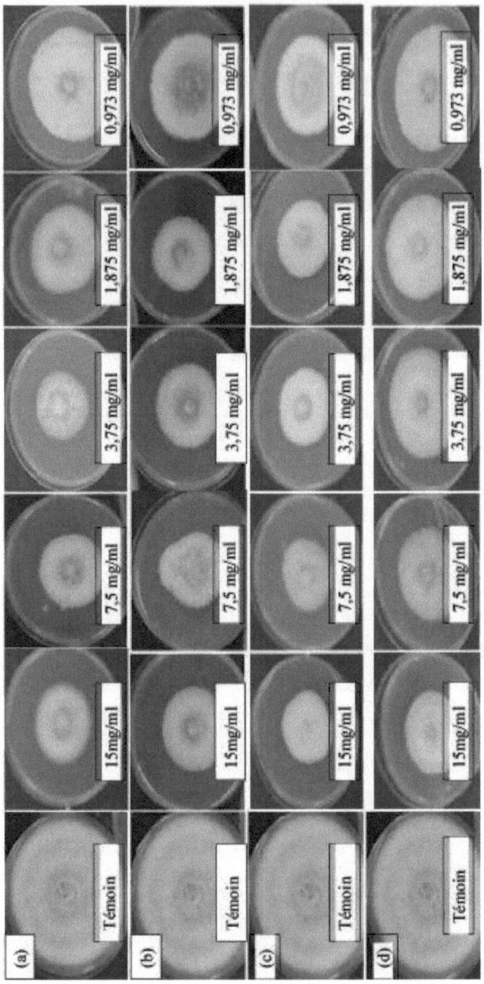

Appendix 05. Inhibition effect of polyphdnolic extracts on Moniliella *sp.* strain (a): PHRS, (b): PHRC, (c): PHBS, (d): PHBC

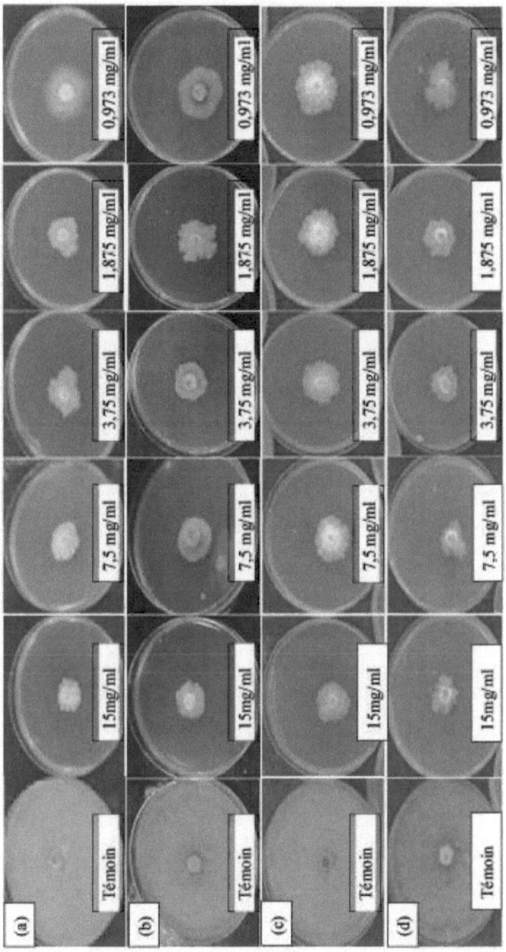

Appendix 06. Inhibition effect of polyphdnolic extracts on the *Rhizopus sp.* strain (a): PEERS, (b): PHRC, (c): PHBS, (d): PHBC.

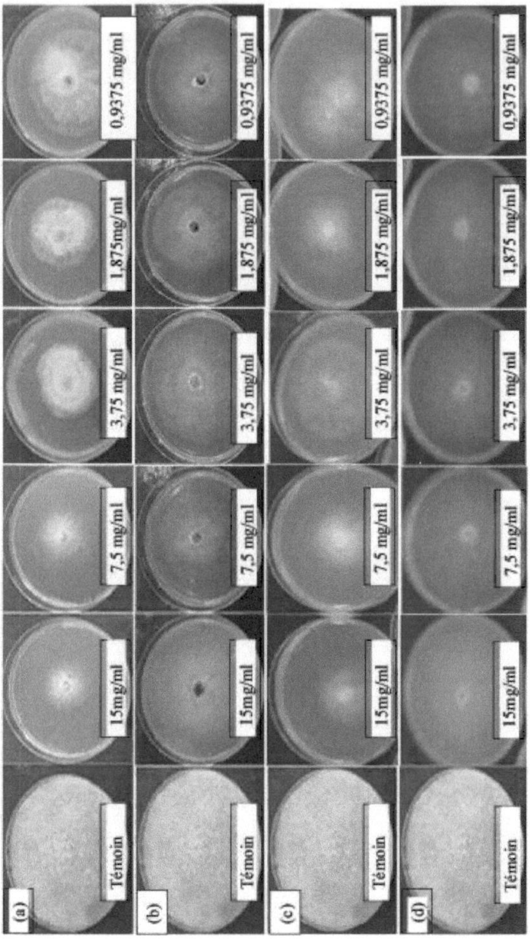

Appendix 07.

A. Inhibition rate of dry bean total polyphënols against *Alternaria.sp.* strain.

Strain	**Concentrations (mg/ml)**	**Inhibition percentages (%)**			
		PHBS	**PHBC**	**PHRS**	**PHRC**
Alternaria sp.	15,00	62,94±0,83	65,29±0,83	63.52±1,66	75,00±0,90
	7,50	61,17±1,66	62,94±0,83	60,00±1,67	69,23±1,81
	3,75	57,64±1,66	62,35±1,66	58,82±0,00	65,38±0,00
	1,87	57,06±5,83	61,76±4,16	56,47±3,32	63,46±4,53
	0,94	56,47±1,67	59,99±6,65	53,53±5,82	62,82±1,81

B. Inhibition rate of dry bean polyphënols against the *Fusarium sp.* strain.

Strain	**Concentrations (mg/ml)**	**Inhibition percentages (%)**			
		PHBS	**PHBC**	**PHRS**	**PHRC**
Fusarium sp.	15,00	49,37±0,88	18,75±1,76	39,37±0,88	45,29±4,16
	7,50	47,50±1,77	16,25±0,00	31,25±0,00	42,35±0,00
	3,75	46,87±4,42	13,75±1,77	30,62±0,88	39,41±2,50
	1,87	45,87±0,88	12,50±0,00	28,75+5,30	34,12+0,00
	0,94	45,00±01,77	10,00±0,00	26,87±0,88	32,35±4,16

C. Inhibition rate of dry bean polyphënols against the *Moniliella sp.* strain.

Strain	**Concentrations (mg/ml)**	**Inhibition percentages**			
		PHBS	PHBC	PHRS	PHRC
Moniliella sp.	15	72,5±0	80±1,77	81,76±0,83	77,06±0,83
	7,5	68,75±0	78,12±4,42	78,23±0,83	75,88±0,83
	3,75	68,75±5,30	75,62±0,88	77,64±1,66	75,29±0
	1,87	65±1,77	75±0	76,47±0	72,94±1,67
	0,94	63,1±2	71,25±1,77	75,32±1,62	68,82±2,50

D. Inhibition rate of dry bean polyphënols against the *Rhizopus sp.* strain.

Strain	**Concentrations (mg/ml)**	**Inhibition percentages**			
		PHBS	**PHBC**	**PHRS**	**PHRC**
Rhizopus sp.	15,00	54,18±0,00	26,47±4,16	59,41±4,16	44,11±2,50
	7,50	33,23±1,34	24,70±1,66	44,70±4,98	42,94±0,83
	3,75	30,00±0,83	22,94±0,83	33,53±5,83	39,41±2,50
	1,87	30,85±2,04	20,00±0,00	32,35±0,83	34,12±0,00
	0,94	28,82±4,16	17,65±0,00	31,76±4,99	32,33±4,13

Resume

The aim of this study was to evaluate the antifungal activity of polyphenols extracted from the whole-grain flour of two varieties of dry bean, one white (*Tima*) and the other red (*MGT djedida*). Determination of the moisture content of the two dry bean varieties revealed levels that were favourable for mould growth. Isolation, purification and microscopic study of the isolated strains identified six genera of moulds: *Alternaria, Aspergillus, Fusarium, Moniliella, Penicillium* and *Rhizopus.* Total polyphenols were extracted using a polar solvent and quantified using the Folin Ciocalteu reaction. The total polyphenol content was 0.40±0.005 mg EAG/g for the red variety and 0.27±0.005 mg EAG/g for the white variety. The antifungal test carried out by the direct contact method and the dilution method, and the antifungal index (IA_{100}) proved the antifungal power of the polyphenols, with *Alternaria sp.*, *Moniliella sp.* and *Rhizopus sp. being the* most sensitive strains. The most active polyphenols are those extracted from the healthy variety of kidney bean. On the basis of this study, we can conclude that the results obtained are promising, encouraging the development of varieties with a high polyphenol content, a natural, bioeffective preservative that can be an alternative solution to chemical control for crop protection in storage facilities.

Key words: Polyphenols, antifungal activity, dry beans.

MIX
Papier aus verantwortungsvollen Quellen
Paper from responsible sources
FSC® C105338

Printed by Books on Demand GmbH, Norderstedt / Germany